W9-BZW-814

Ad Hoc Mobile Wireless Networks

Ad Hoc Mobile Wireless Networks

Principles, Protocols, and Applications

Subir Kumar Sarkar
T G Basavaraju
C Puttamadappa

 Auerbach Publications
Taylor & Francis Group
New York   London

CRC Press is an imprint of the
Taylor & Francis Group, an **informa** business

Auerbach Publications
Taylor & Francis Group
6000 Broken Sound Parkway NW, Suite 300
Boca Raton, FL 33487-2742

International Standard Book Number-13: 978-1-4200-6221-2 (Hardcover)

Library of Congress Cataloging-in-Publication Data

Sarkar, Kumar.
    Ad hoc mobile wireless networks : principles, protocols, and applications / Subir Kumar Sarkar, T. G. Basavaraju, and C. Puttamadappa.
        p. cm.
    Includes bibliographical references and index.
    ISBN 978-1-4200-6221-2 (alk. paper)
    1. Wireless communication systems--Quality control. 2. Internetworking (Telecommunication) I. Basavaraju, T. G. II. Puttamadappa, C. III. Title.

TK5103.2.S27 2007
621.384--dc22   2007020021

Visit the Taylor & Francis Web site at
http://www.taylorandfrancis.com

and the Auerbach Web site at
http://www.auerbach-publications.com

Contents

Preface

Ad hoc networks are autonomous systems which comprise a collection of mobile nodes that use wireless transmission for communication. They are self-organized, self-configured, and self-controlled infrastructure-less networks. This type of network can be set up or deployed anywhere and anytime because it poses very simple infrastructure setup and no or minimal central administration. These networks are mainly used by community users such as military, researchers, business, students, and emergency services. This book addresses and explains network concepts, mechanism, design, and performance. *Ad Hoc Mobile Wireless Networks: Principles, Protocols, and Applications* presents the latest techniques, solutions, and support to understand the concepts easily with suitable examples. The book begins with wireless network fundamentals covering Bluetooth, IrDA, HomeRF, WiFi, WMax, Wireless Internet, and MobileIP.

Coverage includes the following:

- Introduction to mobile ad hoc networks
- MAC layer protocols for ad hoc wireless networks
- Routing protocols for ad hoc wireless networks
- Multicast routing protocols for mobile ad hoc networks
- Transport layer protocols for ad hoc networks
- Quality of service in ad hoc wireless networks
- Energy management in ad hoc wireless networks
- Mobility models for multihop ad hoc wireless networks
- Cross-layer design issues for ad hoc wireless networks
- Recent developments in ad hoc wireless networks

Each topic is explained with suitable examples, and problems are included at the end of each chapter. This book is a very useful resource for researchers, academicians, network engineers, and students designing or developing ad hoc wireless networks.

About the Authors

Subir Kumar Sarkar, Ph.D., is a professor in the Department of Electronics and Telecommunication Engineering, Jadavpur University, Kolkata, West Bengal, India. He completed his B. Tech. and M. Tech. from the Institute of Radio Physics and Electronics, University of Calcutta, Kolkata. He holds a Ph.D. (Tech.) degree from the University of Calcutta in microelectronics. His present fields of interest include the application of soft computing tools in simulation of device models, high-frequency and low-power-consuming devices and their parameter optimization, networking, and mobile communication. He has authored two textbooks, *Optical Fiber and Fiber Optic Communication System* and *Operational Amplifier and Their Applications* (published by S. Chand and Company Private Limited, New Delhi).

T. G. Basavaraju is a professor and head of the Department of Computer Science and Engineering, Sapthagiri College of Engineering, Bangalore, India. He obtained his bachelor's degree in engineering from Kuvempu University and master's degree in engineering from Bangalore University. He has worked for Motorola, Axes Technologies, and Samsung India Software Operations as a software engineer. His areas of interest include performance analysis and evaluation of routing protocols in ad hoc wireless networks.

C. Puttamadappa, Ph.D., is a professor and head of the Department of Electronics and Communication, Manipal Institute of Technology, Manipal, Karnataka, India. He obtained his bachelor's degree in engineering from Mysore University and master's degree in engineering from Bangalore University. He received his Ph.D. from Jadavpur University, Kolkata. His areas of interest include electron devices, ad hoc networks, VLSI, and nanotechnology.

Chapter 1

Introduction

1.1 Fundamentals of Wireless Networks

Communication between various devices makes it possible to provide unique and innovative services. Although this interdevice communication is a very powerful mechanism, it is also a complex and clumsy mechanism, leading to a lot of complexity in the present-day systems. This not only makes networking difficult but limits its flexibility as well. Many standards exist today for connecting various devices. At the same time, every device has to support more than one standard to make it interoperable between different devices. Take the example of setting up a network in an office. Right now, entire office buildings have to make provisions for lengths of cable that stretch kilometers through conduits in the walls, floors, and ceilings to workers' desks.

In the last few years, many wireless connectivity standards and technologies have emerged. These technologies enable users to connect a wide range of computing and telecommunications devices easily and simply, without the need to buy, carry, or connect cables. These technologies deliver opportunities for rapid ad hoc connections and the possibility of automatic, unconscious connections between devices. They will virtually eliminate the need to purchase additional or proprietary cabling to connect individual devices, thus creating the possibility of using mobile data in a variety of applications. Wired local area networks (LANs) have been very successful in the last few years, and now with the help of these wireless connectivity technologies, wireless LANs (WLANs) have started emerging as much more powerful and flexible alternatives to the wired LANs. Until a year ago, the speed of the WLAN was limited to two megabits per second (Mbps), but with the introduction of these new standards, we are seeing WLANs that can support up to eleven Mbps in the Industrial, Scientific, and Medical (ISM) band.

There are many such technologies and standards, and notable among them are Bluetooth, Infrared Data Association (IrDA), HomeRF, and Institute of Electrical and Electronic Engineers (IEEE) 802.11 standards. These technologies compete in certain fronts and are complementary in other areas. So, given the fact that so many technologies exist, which technology is the best, and which solution should one select for a specific application? To be able to understand this, we need to look at the strengths and weaknesses and also the application domains of each of these standards and technologies. The premise behind all these standards is to use some kind of underlying radio technology to enable wireless transmission of data, and to provide support for forming networks and managing various devices by means of high-level software. Bluetooth, though quite new, has emerged as the front-runner in this so-called battle between competing technologies due to the kind of support it is getting from all sections of the industry. However, it must be kept in mind that the viability of a technology depends on the application context.

1.1.1 Bluetooth

Bluetooth is a high-speed, low-power, microwave wireless link technology designed to connect phones, laptops, personal digital assistants (PDAs), and other portable equipment with little or no work by the user. Unlike infrared, Bluetooth does not require line-of-sight positioning of connected units. The technology uses modifications of existing wireless LAN techniques but is most notable for its small size and low cost. Whenever any Bluetooth-enabled devices come within range of each other, they instantly transfer address information and establish small networks between each other, without the user being involved.

Features of Bluetooth technology are as follows:

- Operates in the 2.56 gigahertz (GHz) ISM band, which is globally available (no license required)
- Uses Frequency Hop Spread Spectrum (FHSS)
- Can support up to eight devices in a small network known as a "piconet"
- Omnidirectional, nonline-of-sight transmission through walls
- 10 m to 100 m range
- Low cost
- 1 mw power
- Extended range with external power amplifier (100 meters)

1.1.2 IrDA

IrDA is an international organization that creates and promotes interoperable, low-cost, infrared data interconnection standards. IrDA has a set of protocols covering all layers of data transfer and, in addition, has some network management and interoperability designs. IrDA protocols have IrDA DATA as the vehicle for data delivery and IrDA CONTROL for sending the control information. In general, IrDA is used to provide wireless connectivity technologies for devices that would normally use cables for connectivity. IrDA is a point-to-point, narrow-angle (30° cone), ad hoc data transmission standard designed to operate over a distance of zero to one meter and at speeds of 9600 bits per second (bps) to 16 Mbps. Adapters now include the traditional upgrades to serial and parallel ports.

Features of IrDA are as follows:

- Range: From contact to at least one meter, and can be extended to two meters. A low-power version relaxes the range objective for operation from contact through at least 20 centimeters (cm) between low-power devices and 30 cm between low-power and standard-power devices. This implementation affords ten times less power consumption.
- Bidirectional communication is the basis of all specifications.
- Data transmission from 9600 bps with primary speed or cost steps of 115 kilobits per second (kbps) and maximum speed of up to 4 Mbps.
- Data packets are protected using a Cyclic Redundancy Check (CRC) (CRC-16 for speeds up to 1.152 Mbps, and CRC-32 at 4 Mbps).

1.1.2.1 Comparison of Bluetooth and IrDA

Bluetooth and IrDA are both critical to the marketplace. Each technology has advantages and drawbacks, and neither can meet all users' needs. Bluetooth's ability to penetrate solid objects and its capability for maximum mobility within the piconet allow for data exchange applications that are very difficult or impossible with IrDA. For example, with Bluetooth, a person could synchronize his or her phone with a personal computer (PC) without taking the phone out of a pocket or purse; this is not possible with IrDA. The omnidirectional capability of Bluetooth allows synchronization to start when the phone is brought into range of the PC.

On the other hand, in applications involving one-to-one data exchange, IrDA is at an advantage. Consider an application where there are many people sitting across a table in a meeting. Electronic cards can be exchanged between any two people by pointing their IrDA devices toward each other (because of the directional nature). In contrast, because Bluetooth is omnidirectional in nature, the Bluetooth device will detect all similar devices in the room and the user would have to select the intended person from, say, a list provided by the Bluetooth device. On the security front, Bluetooth provides security mechanisms which are not present in IrDA.

However, the narrow beam (in the case of IrDA) provides a low level of security. IrDA beats Bluetooth on the cost front. The Bluetooth standard defines layers 1 and 2 of the Open System Interconnection (OSI) model. The application framework of Bluetooth is aimed to achieve interoperability with IrDA and Wireless Access Protocol (WAP). In addition, a host of other applications will be able to use the Bluetooth technology and protocols.

1.1.3   HomeRF

HomeRF is a subset of the International Telecommunication Union (ITU) and primarily works on the development of a standard for inexpensive radio frequency (RF) voice and data communication. The HomeRF Working Group has also developed the Shared Wireless Access Protocol (SWAP). SWAP is an industry specification that permits PCs, peripherals, cordless telephones, and other devices to communicate voice and data without the use of cables. SWAP is similar to the Carrier Sense Multiple Access with Collision Avoidance (CSMA/CA) protocol of IEEE 802.11 but with an extension to voice traffic. The SWAP system can operate either as an ad hoc network or as an infrastructure network under the control of a connection point. In an ad hoc network, all stations are peers, and control is distributed between the stations and supports only data. In an infrastructure network, a connection point is required so as to coordinate the system, and it provides the gateway to the public switched telephone network (PSTN). Walls and floors do not cause any problems in its functionality, and some security is also provided through the use of unique network IDs. It is robust and reliable, and minimizes the impact of radio interference.

Features of HomeRF are as follows:

- Operates in the 2.45 GHz range of the unlicensed ISM band.
- Range: up to 150 feet.
- Employs frequency hopping at 50 hops per second.
- It supports both a Time Division Multiple Access (TDMA) service to provide delivery of interactive voice and a CSMA/CA service for delivery of high-speed data packets.
- The network is capable of supporting up to 127 nodes.
- Transmission power: 100mW.
- Data rate: 1 Mbps using 2 frequency-shift keying (FSK) modulation and 2 Mbps using 4 FSK modulation.
- Voice connections: up to 6 full duplex conversations.
- Data security: blowfish encryption algorithm (over 1 trillion codes).
- Data compression: Lempel-Ziv Ross Williams 3 (LZRW3)-A Algorithm.

Table 1.1 Comparison of Various Wireless Technologies

	Peak Data Rate	*Range*	*Relative Cost*	*Voice Network Support*	*Data Network Support*
IEEE 802.11	2 Mbps	50 meters	Medium	Via Internet Protocol (IP)	Transmission Control Protocol (TCP)/IP
IrDA	16 Mbps	< 2 meters	Low	Via IP	Via Point-to-Point Protocol (PPP)
Bluetooth	1 Mbps	< 10 meters	Medium	Via IP and Cellular	Via PPP
HomeRF	1.6 Mbps	50 meters	Medium	Via IP and PSTN	TCP/IP

1.1.3.1 Comparison of Bluetooth with Shared Wireless Access Protocol (SWAP)

Currently SWAP has a larger installed base compared to Bluetooth, but it is believed that Bluetooth is eventually going to prevail. Bluetooth is a technology to connect devices without cables. The intended use is to provide short-range connections between mobile devices and to the Internet via bridging devices to different networks (wired and wireless) that provide Internet capability. HomeRF SWAP is a wireless technology optimized for the home environment. Its primary use is to provide data networking and dial tones between devices such as PCs, cordless phones, Web tablets, and a broadband cable or Digital Subscriber Line (DSL) modem. Both technologies share the same frequency spectrum but do not interfere with each other when operating in the same space. As far as comparison with IrDA is concerned, SWAP is closer to Bluetooth in its scope and domain, so the comparison between Bluetooth and IrDA holds good to a large extent between these two also. Comparisons of these technologies are given in Table 1.1.

Wireless networks use finite resources, and a given geographical area with many wireless networks will degrade in performance as more users come on. For example, a building with 20 competing networks can cause interference and slow performance for all users. Wireless networks are flexible and can be deployed quickly using inexpensive radio equipment and antennas. The flexibility of being able to rapidly deploy a network means that many networks operating in the same area can "peer," or aggregate themselves into a larger network with more capacity to be used by users. Wireless networks act in a similar manner to people discussing something in a public area. The discussion can be "heard" by others in the area with appropriate equipment. Security issues are thus pushed to the users, forcing the use of encryption and "safe computing" practices that are generally avoided by the public at large today. Wireless network speeds do not (yet) fare well against the gigabit speeds achieved by wired networks such as gigabit Ethernet or Fiber. However, wireless network technology is rapidly maturing, and new, open standards are emerging that will provide speeds comparable to those of Fiber and other infrastructures. Wireless network technologies based on IEEE 802.11 and 802.16 standards (wireless fidelity [WiFi] and Worldwide Interoperability for Microwave Access [WiMax]) are not restricted to any one vendor and can be deployed by anyone with a basic understanding of the technology. Wireless networks are ideal for connecting many people without the expenses of deploying cable and human resources. Wireless networks provide mobility and access to information based on physical proximity.

A typical wireless network consists of

- an access point, and
- client wireless radios used by each subscriber.

The access point is a "central hub" device that provides service to 1–100 subscribers. Multiple access points may be required in larger geographic areas or to serve large groups of users. An access point can be connected to other access points or connected directly to the network that provides the connection to the Internet in one's community. The access point is typically placed in a central location within view of a group of subscribers and within view of other access points or with a network link to a Point Of Presence (POP).

The access point manages the flow of information between subscribers and to other elements in the network. It broadcasts a network Service Set ID (SSID), or network name, and handles limited security functions. When a subscriber links to the community wireless network, his or her subscriber radio is configured to use the access point's SSID and relevant security parameters. The subscriber radio then establishes a connection to the wireless network, and a data connection is created.

A computer system is connected to a wireless device using an Ethernet cable. Information sent from the computer (or other computers on the same Ethernet network) are delivered to the wireless device:

- A transmitter sends radio signals with information to an antenna.
- The antenna takes the radio signals, directs them into the air, and directs them toward a specific physical location.
- A receiver hears the radio signals by way of its own antenna, and converts them into a format that the user's computer can use.

Once the radio signal leaves the transmitter's antenna, it travels through the air and is picked up by receiving antennas. As the signal travels through the air, it loses its strength, eventually losing enough power so that it cannot be accurately received.

Wireless networks take many forms. VHF radio, FM–AM radio, cellular phones, and CB radios are all forms of wireless technology but have very specific purposes (usually for the purpose of communicating verbal information). When one talks about wireless networking, it is about a breed of technology that is able to communicate data. Data can be voice, the Internet, or any other kind of computer information. This kind of wireless technology can be used to supplement or even replace existing wireless systems.

There are many wireless technologies suitable for data networking. When the concept of using radio signals to connect various computers in a building was introduced, the IEEE formed a committee to set the standards for the technology. That committee was called the 802.11 committee, and the various standards they developed are known as 802.11a, or 802.11b, 802.11g, and so forth. This group of 802.11 standards became known as WiFi technology. Because WiFi technology quickly became popular, the cost of WiFi equipment has decreased rapidly. Many organizations and wireless Internet Service Providers (ISPs) have started with WiFi.

1.1.4 802.11 (WiFi)

WiFi is a common wireless technology used by home owners, small businesses, and starting ISPs. WiFi devices are available "off the shelf" from computer stores, and enhanced WiFi devices are designed for ISP use.

Advantages of WiFi are as follows:

■ Ubiquitous and vendor neutral; any WiFi device will work with another regardless of the manufacturer.
■ Affordable cost.
■ Hackable; many "hacks" exist to extend the range and performance of a WiFi network.

Disadvantages are as follows:

■ Designed for LANs, not wide area networking (WAN).
■ Uses the CSMA mechanism. Only one wireless station can "talk" at a time, meaning one user can potentially hog all of the network's resources. Applications such as video conferencing, Voice-Over Internet Protocol (VOIP), and multimedia can take down a network.

1.1.5 802.16 (WiMax)

WiMax is a superset of WiFi and is designed specifically for last-mile distribution and mobility. WiMax promises high speed (30 Mbps+). WiMax is a relatively new standard; thus, WiMax products are expensive.

An advantage of WiMax is as follows:

■ Specifically designed for wide area networking.

Disadvantages of WiMax are the following:

■ New technology; has not passed the test of time (yet).
■ More expensive than WiFi.

1.1.6 Hotspots

Hotspots are wireless networks often run by businesses and individuals. They are called "hotspots" because they provide a small coverage area for people to connect to community networks and the Internet; popular locations for hotspots include communal areas such as restaurants and cafés.

Hotspots are also powerful tools for supporting tourism. Visitors to a hotspot can be presented with information about the local community, including upcoming events and even presentations of local artwork and artisan works. The BC Wireless Network Society of British Columbia, Canada, provides a service for a Community Wireless Hotspot Network.

1.1.7 Mesh Networking

Mesh networking is the holy grail of wireless networking. "Mesh" refers to many types of technology that enable wireless systems to automatically find each other and self-configure themselves to route information amongst themselves.

Mesh is as organic as networks can get, but it is very immature. Several implementations exist (but are not compatible with each other). Mesh networking should be treated as experimental, but community wireless networks make provisions for using mesh technology either during early deployment (where it may turn out to be stable for the needs of the community) or on an experimental basis. Most mesh products work under the Linux operating system and can use Prism 2.0 and 2.5 devices, or Atheros-based radios.

Some popular mesh protocols that exist are as follows:

■ AODV: An older protocol used by commercial and open source products such as LocustWorld. AODV appears to have many flaws, and is not *necessarily* recommended.
■ RoofNet: An experimental protocol from MIT. RoofNet is being tested by community wireless networks throughout the world, and appears to be very promising.

1.1.7.1   Limitation of Wireless Technology

The wireless radio spectrum is a finite resource. Many people can use the radio spectrum, but as more people use wireless networking, interference will increase. In some cases you may even find your competitors actively working to interfere with you. It is important to adopt a policy early on in your network deployment to work with your community to resolve interference issues. Network operators should inform each other when setting up a new wireless system. In fact, if you use licensed wireless devices, you must coordinate with other wireless users. Although coordination is not required when using license-exempt wireless devices, it is a best practice to follow.

1.2   Wireless Internet

Wireless Internet has become possible through the evolution of portable computers and wireless connections over a mobile telephone network. However, the realization of a mobile computing environment requires a communication architecture which not only is compatible with the current architectures but also takes into account the specific features of mobility and wirelessness.

In the last few years, we have seen an increase in the use of Internet systems as well as an increase in mobile communications. Now, many services of high utility to the end users are based on the Internet technology. If a convergence of the mobile and Internet technologies can be achieved, it would be powerful in realizing vast economies of scale as well as highly flexible services platforms. But, to manage a reliable wireless Internet, three kinds of constraints have to be studied:

- The wireless operating environment
- The existing Internet architecture
- The limitation of the end devices

Wireless networks are very interesting for the following reasons:

- Mobility
- Reduced installation time
- Increased reliability
- Long-term cost savings

The Internet is a cooperatively run collection of computer networks that span the globe. It is also a vast collection of resources: people, information, and multimedia. The word "Internet" describes a number of agreements, arrangements, and connections. In fact, it is a network of networks—more precisely, a network of local area

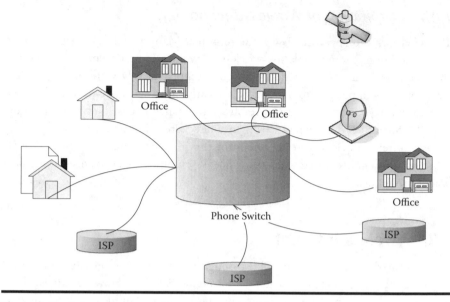

Figure 1.1 Internet connection.

networks. Each individual network has its own domain and has specific resources and capabilities. Figure 1.1 shows a simple Internet connection.

The Internet offers a variety of services such as e-mail, keyboard-to-keyboard chatting, real-time voice and video communication, and the transfer, storage, and retrieval of files. The Internet uses a system of packet switching for data transfer. The Internet was designed to be highly robust. In case one section of the network became inoperable, packets could simply be sent over another route and reach their destination. An important part of the IP protocol is the IP addressing standards, which define mechanisms to provide a unique address for each computer on the Internet. Users connect to an ISP via modems or Integrated Service Digital Networks (ISDN), and the ISP routes the Transmission Control Protocol (TCP)/IP packets to and from the Internet.

The characteristics of wireless networks showed us that to manage reliable wireless Internet, we definitely have to consider the following subjects:

- Speed of wireless link
- Scalability
- Mobility
- Limited battery power
- Disconnection (voluntary or involuntary)
- Replication caching
- Handover

1.2.1 IP Limitations

The IP has limitations due to its proper characteristics:

■ To send a packet on the Internet, a computer must have an IP address.
■ This IP address is associated with the computer's physical location.
■ TCP/IP protocol routes packets to their destination according to the IP address.

That leads to a big limitation. Indeed, within TCP/IP, if the mobile user moves without changing the IP address, the routing is lost; if the user changes the IP address, connections are lost. In both cases, packets are lost. That leads to an unreliable network.

Regarding the specific features of mobility and wirelessness, wireless Internet must do the following:

■ Give mobile users the full Internet experience, not just a limited menu of specialized Web services, or only e-mail.
■ Indeed, voice telephony should migrate to the wireless Internet in due time.
■ Be reasonably fast: at least 100,000-bps throughput per user, about what has proved commercially successful over dial-up lines, with a growth path to millions of bits per second.
■ Work indoors and outdoors to both stationary and mobile users. (Although drivers should not be surfing the Web, they may listen to Internet radio stations.)
■ Use power efficiently, because most devices will run on batteries or fuel cells for at least a few hours on a single charge.
■ Scale up to support millions of active devices, or more, within a single metropolitan region.

1.2.2 Mobile Internet Protocol (IP)

Mobile IP is an emerging set of protocols created by the Internet Engineering Task Force (IETF). Basically, it is a modification to IP that allows nodes to continue to receive packets, independently of their connection point to the Internet. Figure 1.2 shows a mobile node communicating with other nodes after changing its link-layer point of attachment to the Internet, and it does not change its IP address. However, mobile IP is not suitable for fast mobility and smooth handover between cells, and a few requirements are to be considered for its design.

The messages used to transmit information about the location of a mobile node to another node must be authenticated to protect against remote redirection attacks.

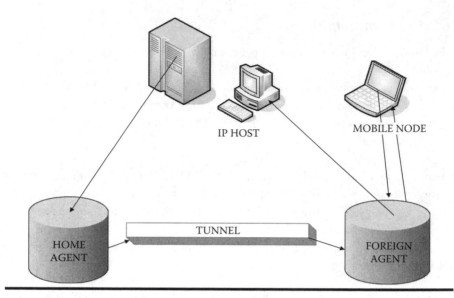

Figure 1.2   Datagram routing using Mobile IP.

1.2.2.1   Working of Mobile IP

IP routes packets from a source endpoint to a destination by allowing routers to forward packets from incoming network interfaces to outbound interfaces according to information available in the routing tables. The routing tables typically maintain the next-hop information for each destination IP address, according to the number of networks to which that IP address is connected. The network number is derived from the IP address by masking off some of the low-order bits. Thus, the IP address typically carries with it information that specifies the IP node's point of attachment.

To maintain existing transport-layer connections as the mobile node moves from place to place, it must keep its IP address the same. In TCP, connections are indexed by a quadruplet that contains the IP addresses and port numbers of both connection endpoints. Changing any of these four numbers will cause the connection to be disrupted and lost. On the other hand, the correct delivery of packets to the mobile node's current point of attachment depends on the network number contained within the mobile node's IP address, which changes at new points of attachment. To change the routing requires a new IP address associated with the new point of attachment.

Mobile IP has been designed to solve this problem by allowing the mobile node to use two IP addresses. In Mobile IP, the home address is static and is used, for instance, to identify TCP connections. The care-of address changes at each new point of attachment and can be thought of as the mobile node's topologically significant address; it indicates the network number and thus identifies the mobile node's point of attachment with respect to the network topology. The home address

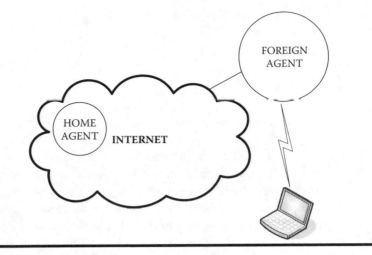

Figure 1.3 Mobile IP agents.

makes it appear that the mobile node is continually able to receive data on its home network, where Mobile IP requires the existence of a network node known as the home agent. Whenever the mobile node is not attached to its home network (and is therefore attached to what is termed a "foreign network"), the home agent gets all the packets destined for the mobile node and arranges to deliver them to the mobile node's current point of attachment. See Figure 1.3.

Whenever the mobile node moves, it registers its new care-of address with its home agent. To get a packet to a mobile node from its home network, the home agent delivers the packet from the home network to the care-of address. The further delivery requires that the packet be modified so that the care-of address appears as the destination IP address. This modification can be understood as a packet transformation or, more specifically, a redirection. When the packet arrives at the care-of address, the reverse transformation is applied so that the packet once again appears to have the mobile node's home address as the destination IP address. When the packet arrives at the mobile node, addressed to the home address, it will be processed properly by TCP or whatever higher level protocol logically receives it from the mobile node's IP (that is, layer-3) processing layer.

In Mobile IP the home agent redirects packets from the home network to the care-of address by constructing a new IP header that contains the mobile node's care-of address as the destination IP address. This new header then shields or encapsulates the original packet, causing the mobile node's home address to have no effect on the encapsulated packet's routing until it arrives at the care-of address. Such encapsulation is also called "tunneling," which suggests that the packet burrows through the Internet, bypassing the usual effects of IP routing.

A mobile node should minimize the number of administrative messages. Mobile IP must place no additional constraints on the assignment of IP addresses.

The Mobile IP can be described with the following steps:

- Step 1: Agent discovery
- Step 2: Registration home agent
- Step 3: Tunneling

A mobile node operating away from home registers its new care-of address with its home agent through the exchange of a registration request and registration reply messages. The home agent tunnels the information packets to the care-of-address when the mobile node is away. Packets sent to the mobile node's home address are intercepted by its home agent, which tunnels them to the appropriate care-of address. There, the packets are delivered to the mobile node. In the reverse direction, packets sent by the mobile node may be delivered to their destination using a standard IP routing scheme, without necessarily passing through the home agent.

Mobile IP enables mobile computers to move about the Internet but remain addressable via their home network. Each mobile computer has an IP address (a home address) on its home network. Datagrams arriving for the mobile computer at its home network are subsequently repackaged for delivery to the mobile computer at its care-of address. The mobile computer informs its home agent about its current care-of address, using Mobile IP registration protocols. When the home agent receives the mobile computer's care-of address, it binds that address with the mobile computer's home address, forming a binding that has an associated life-time of validity. This process is called "registration," and is transacted between the mobile computer and the home agent each time the mobile computer changes its point of attachment and receives a new care-of address. Often, the care-of address is advertised by an entity known as a foreign agent, which is located nearby the mobile computer and relays the registration messages back and forth between the mobile node and the home agent. Other times, the mobile computer itself acquires a care-of address by other means (notably, via the Dynamic Host Configuration Protocol [DHCP]) and assigns that care-of address to one of its own interfaces. This configuration is known as a "colocated care-of address."

A mobile computer can easily switch between the two modes of operation depending upon the way care-of addresses are provided at its various points of attachment. Figure 1.4 shows a thumbnail sketch of a typical configuration, where the foreign agent has advertised the care-of address used by the mobile computer; the foreign agent and home agent are presumably and typically located on different subnets which have no a priori relationship to each other. If the mobile computer had attached via DHCP, there would be no foreign agent, but there would still be (typically) no relationship between the home network and the new point of attachment of the mobile computer.

When a home agent has a valid binding for the mobile node and a datagram for the mobile computer arrives at the home network, the home agent receives the

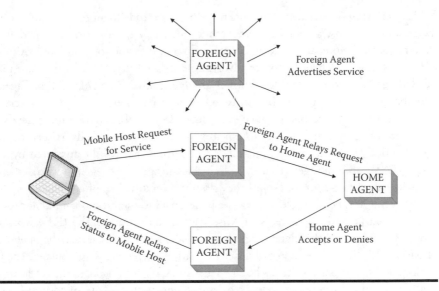

Figure 1.4 Registration operations in Mobile IP.

datagram, acting as a proxy agent for the mobile computer on the home network. The home agent subsequently tunnels (by encapsulation) the datagram to the mobile computer's care-of address. The tunnel is the path between the home agent and the care-of address, and the care-of address is also known as the tunnel endpoint. After the datagram arrives at the tunnel endpoint, it is decapsulated and final delivery is made to the mobile computer. When the mobile node has a colocated care-of address, then the final delivery is accomplished trivially.

Because traffic to the mobile node is controlled by correct operation of the Mobile IP registration protocol, it is of essential importance that no corruption or intentional modifications of registration message data go undetected. If a malicious agent were able to register its own IP address as a false care-of address for the mobile node, the home agent would then route all the datagrams for the mobile node to the malicious agent instead. Clearly, the home agent must be able to ascertain that registration messages were issued authentically by the mobile node itself. This is accomplished by affixing a 128-bit digital signature, computed by using Message-Digest algorithm 5 (MD5) as a one-way hash function, to the registration messages, and including protection against replay attacks, in which a malicious node could record valid registrations for later replay, effectively disrupting the ability of the home agent to tunnel to the current care-of address of the mobile node at that later time.

1.2.3 Discovering the Care-of Address

The Mobile IP discovery process has been built on top of an existing standard protocol, Router Advertisement. Mobile IP discovery does not modify the original fields of existing router advertisements but simply extends them to associate mobility functions. Thus, a router advertisement can carry information about default routers, just as before, and in addition carry further information about one or more care-of addresses. When the router advertisements are extended to also contain the needed care-of address, they are known as "agent advertisements." Home agents and foreign agents typically broadcast agent advertisements at regular intervals (for example, once a second or once every few seconds). If a mobile node needs to get a care-of address and does not wish to wait for the periodic advertisement, the mobile node can broadcast or multicast a solicitation that will be answered by any foreign agent or home agent that receives it. Home agents use agent advertisements to make themselves known, even if they do not offer any care-of addresses.

However, it is not possible to associate preferences to the various care-of addresses in the router advertisement, as is the case with default routers. The IETF working group was concerned that dynamic preference values might destabilize the operation of Mobile IP. Because no one could defend static preference assignments except for backup mobility agents, which do not help distribute the routing load, the group eventually decided not to use the preference assignments with the care-of address list.

Thus, an agent advertisement performs the following functions:

- Allows for the detection of mobility agents.
- Lists one or more available care-of addresses.
- Informs the mobile node about special features provided by foreign agents, for example, alternative encapsulation techniques.
- Lets mobile nodes determine the network number and status of their link to the Internet.
- Lets the mobile node know whether the agent is a home agent, a foreign agent, or both, and therefore whether it is on its home network or a foreign network.

Mobile nodes use router solicitations to detect any change in the set of mobility agents available at the current point of attachment. (In Mobile IP, this is then termed "agent solicitation.") If advertisements are no longer detectable from a foreign agent that previously had offered a care-of address to the mobile node, the mobile node should presume that the foreign agent is no longer within range of the mobile node's network interface. In this situation, the mobile node should begin to hunt for a new care-of address, or possibly use a care-of address known from advertisements it is still receiving. The mobile node may choose to wait for another advertisement if it has not received any recently advertised care-of addresses, or it may send an agent solicitation.

1.2.4 Registering the Care-of Address

Once a mobile node has a care-of address, its home agent must find out about it. Figure 1.4 shows the registration process defined by Mobile IP for this purpose.

The process begins when the mobile node, possibly with the assistance of a foreign agent, sends a registration request with the care-of address information. When the home agent receives this request, it (typically) adds the necessary information to its routing table, approves the request, and sends a registration reply back to the mobile node. Although the home agent is not required by Mobile IP to handle registration requests by updating entries in its routing table, doing so offers a natural implementation strategy.

1.2.5 Authentication

Registration requests contain parameters and flags that characterize the tunnel through which the home agent will deliver packets to the care-of address. Tunnels can be constructed in various ways. When a home agent accepts the request, it begins to associate the home address of the mobile node with the care-of address, and maintains this association until the registration lifetime expires. The triplet that contains the home address, care-of address, and registration lifetime is called a "binding" for the mobile node. A registration request can be considered a "binding update" sent by the mobile node.

To secure the registration request, each request must contain unique data so that two different registrations will in practical terms never have the same MD5 hash. Otherwise, the protocol would be susceptible to replay attacks. To ensure this does not happen, Mobile IP includes within the registration message a special identification field that changes with every new registration. The exact semantics of the identification field depend on several details, which are described at greater length in the protocol specification. Briefly, there are two main ways to make the identification field unique. One is to use a time stamp; then, each new registration will have a later time stamp and thus differ from previous registrations. The other is to cause the identification to be a pseudorandom number; with enough bits of randomness, it is highly unlikely that two independently chosen values for the identification field will be the same. When randomness is used, Mobile IP defines a method that protects both the registration request and reply from replay, and calls for 32 bits of randomness in the identification field. If the mobile node and the home agent get too far out of synchronization for the use of time stamps, or if they lose track of the expected random numbers, the home agent will reject the registration request and include information to allow resynchronization within the reply.

Using random numbers instead of time stamps avoids problems stemming from attacks on the Network Time Protocol (NTP) that might cause the mobile node to lose time synchronization with the home agent or to issue authenticated registration requests for some future time that could be used by a malicious node to subvert a future registration. The identification field is also used by the foreign agent to match pending registration requests to registration replies when they arrive at the home agent and to subsequently be able to relay the reply to the mobile node. The foreign agent also stores other information for pending registrations, including the mobile node's home address, the mobile node's Media Access Control (MAC) Layer address, the source port number for the registration request from the mobile node, the registration lifetime proposed by the mobile node, and the home agent's address. The foreign agent can limit registration lifetimes to a configurable value that it puts into its agent advertisements. The home agent can reduce the registration lifetime, which it includes as part of the registration reply, but it can never increase it.

1.2.6 Automatic Home Agent Discovery

When the mobile node cannot contact its home agent, Mobile IP has a mechanism that lets the mobile node try to register with another unknown home agent on its home network. The method of automatic home agent discovery works by using a broadcast IP address instead of the home agent's IP address as the target for the registration request. When the broadcast packet gets to the home network, other home agents on the network will send a rejection to the mobile node; however, their rejection notice will contain their address for the mobile node to use in a freshly attempted registration message. The broadcast is not an Internet-wide broadcast, but a directed broadcast that reaches only IP nodes on the home network.

1.2.7 Tunneling to the Care-of Address

Figure 1.5 shows the tunneling operations in Mobile IP. The default encapsulation mechanism that must be supported by all mobility agents using Mobile IP is IP-within-IP. Using IP-within-IP, the home agent, the tunnel source, inserts a new IP header, or tunnel header, in front of the IP header of any datagram addressed to the mobile node's home address. The new tunnel header uses the mobile node's care-of address as the destination IP address, or tunnel destination. The tunnel source IP address is the home agent, and the tunnel header uses 4 as the higher level protocol number, indicating that the next protocol header is again an IP header. In IP-within-IP, the entire original IP header is preserved as the first part of the payload of the tunnel header. Therefore, to recover the original packet, the foreign agent merely has to eliminate the tunnel header and deliver the rest to the mobile node.

Figure 1.5 shows that sometimes the tunnel header uses protocol number 55 as the inner header. This happens when the home agent uses minimal encapsulation instead of IP-within-IP. Processing for the minimal encapsulation header is slightly more complicated than for IP-within-IP, because some of the information from the tunnel header is combined with the information in the inner minimal encapsulation header to reconstitute the original IP header. On the other hand, header overhead is reduced.

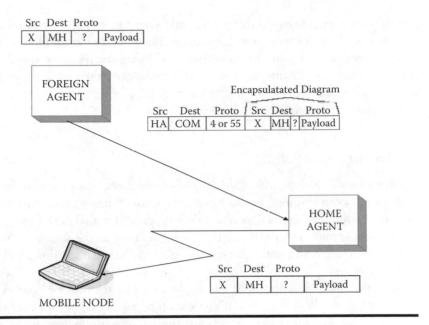

Figure 1.5 Tunneling operations in Mobile IP.

1.2.8 Issues in Mobile IP

The most pressing outstanding problem facing Mobile IP is that of security, but other technical as well as practical obstacles to deployment exist.

1.2.8.1 Routing Inefficiencies

The base Mobile IP specification has the effect of introducing a tunnel into the routing path followed by packets sent by the correspondent node to the mobile node. Packets from the mobile node, on the other hand, can go directly to the correspondent node with no tunneling required. This asymmetry is captured by the term "triangle routing," where a single leg of the triangle goes from the mobile node to the correspondent node, and the home agent forms the third vertex, controlling the path taken by data from the correspondent node to the mobile node.

1.2.8.2 Security Issues

A great deal of attention is being focused on making Mobile IP coexist with the security features coming into use within the Internet. Firewalls, in particular, cause difficulty for Mobile IP because they block all classes of incoming packets that do not meet specified criteria. Enterprise firewalls are typically configured to block packets from entering via the Internet that appear to emanate from internal computers. Although this permits management of internal Internet nodes without great attention to security, it presents difficulties for mobile nodes wishing to communicate with other nodes within their home enterprise networks. Such communications, originating from the mobile node, carry the mobile node's home address, and would thus be blocked by the firewall. Mobile IP can be viewed as a protocol for establishing secure tunnels.

1.2.8.3 Ingress Filtering

Complications are also presented by ingress-filtering operations. Many border routers discard packets coming from within the enterprise if the packets do not contain a source IP address configured for one of the enterprise's internal networks. Because mobile nodes would otherwise use their home address as the source IP address of the packets they transmit, this presents difficulty. Solutions to this problem in Mobile IPv4 typically involve tunneling outgoing packets from the care-of address, but then the difficulty is how to find a suitable target for the tunneled packet from the mobile node. The only universally agreed on possibility is the home agent, but that target introduces yet another serious routing anomaly for communications between the mobile node and the rest of the Internet.

1.2.8.4 User Perceptions of Reliability

The design of Mobile IP is founded on the premise that connections based on TCP should survive cell changes. However, opinion is not unanimous on the need for this feature. Many people believe that computer communications to laptop computers are sufficiently bursty so that there is no need to increase the reliability of the connections supporting the communications. The analogy is made to fetching Web pages by selecting the appropriate URLs. If a transfer fails, people are used to trying again. This is tantamount to making the user responsible for the retransmission protocol and depends for its acceptability on a widespread perception that computers and the Internet cannot be trusted to do things right the first time.

1.2.8.5 Issues in IP Addressing

Mobile IP creates the perception that the mobile node is always attached to its home network. This forms the basis for the reachability of the mobile node at an IP address that can be conventionally associated with its Fully Qualified Domain Name (FQDN). If the FQDN is associated with one or more other IP addresses, perhaps dynamically, then those alternative IP addresses may deserve equal standing with the mobile node's home address. Moreover, it is possible that such an alternative IP address would offer a shorter routing path if, for instance, the address were apparently located on a physical link nearer to the mobile node's care-of address, or if the alternative address were the care-of address itself. Finally, many communications are short-lived and depend on neither the actual identity of the mobile node nor its FQDN, and thus do not take advantage of the simplicity afforded by use of the mobile node's home address. These issues surrounding the mobile node's selection of an appropriate long-term (or not-so-long-term) address for use in establishing connections are complex and are far from being resolved.

1.2.8.6 Slow Growth in the Wireless Local Area Network (LAN) Market

Mobile IP has been engineered as a solution for wireless LAN location management and communications, but the wireless LAN market has been slow to develop. It is difficult to make general statements about the reasons for this slow development, but with the recent ratification of the IEEE 802.11 MAC protocol, wireless LANs may become more popular. Moreover, the bandwidth for wireless devices has been constantly improving, so that radio and infrared devices on the market today offer multimegabyte-per-second data rates. Faster wireless access over standardized MAC layers could be a major catalyst for growth of this market.

1.2.8.7 Competition from Other Protocols

Mobile IP may well face competition from alternative tunneling protocols such as PPTP and L2TP. These other protocols, based on Point-to-Point Protocol (PPP), offer at least portability to mobile computers. If these alternative methods are made widely available, it is unclear if the use of Mobile IP will be displaced or instead made more immediately desirable as people experience the convenience of mobile computing. In the future, it is also possible that Mobile IP could specify use of such alternative tunneling protocols to capitalize on their deployment on platforms that do not support IP-within-IP encapsulation.

1.3 What Are Ad Hoc Networks?

An ad hoc network is a collection of wireless mobile nodes (or routers) dynamically forming a temporary network without the use of any existing network infrastructure or centralized administration. The routers are free to move randomly and organize themselves arbitrarily; thus, the network's wireless topology may change rapidly and unpredictably. Such a network may operate in a stand-alone fashion, or may be connected to the Internet. Multihop, mobility, large network size combined with device heterogeneity, bandwidth, and battery power constraints make the design of adequate routing protocols a major challenge. Some form of routing protocol is in general necessary in such an environment, because two hosts

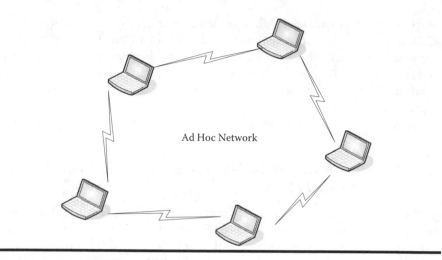

Figure 1.6 Mobile ad hoc network.

that may wish to exchange packets might not be able to communicate directly, as shown in Figure 1.6.

Mobile users will want to communicate in situations in which no fixed wired infrastructure is available. For example, a group of researchers en route to a conference may meet at the airport and need to connect to the wide area network, students may need to interact during a lecture, or firefighters need to connect to an ambulance en route to an emergency scene. In such situations, a collection of mobile hosts with wireless network interfaces may form a temporary network without the aid of any established infrastructure or centralized administration. Because nowadays many laptops are equipped with powerful CPUs, large hard disk drives, and good sound and image capabilities, the idea of forming a network among these researchers, students, or members of a rescue team, who can easily be equipped with the devices mentioned above, seems possible. Such networks received considerable attention in recent years in both commercial and military applications, due to the attractive properties of building a network on the fly and not requiring any preplanned infrastructure such as a base station or central controller.

A mobile ad hoc network (MANET) group has been formed within IETF. The primary focus of this working group is to develop and evolve MANET specifications and introduce them to the Internet standard track. The goal is to support mobile ad hoc networks with hundreds of routers and solve challenges in this kind of network. Some challenges that ad hoc networking faces are limited wireless transmission range, hidden terminal problems, packet losses due to transmission errors, mobility-induced route changes, and battery constraints. Mobile ad hoc networks could enhance the service area of access networks and provide wireless connectivity into areas with poor or previously no coverage (e.g., cell edges). Connectivity to wired infrastructure will be provided through multiple gateways with possibly different capabilities and utilization. To improve performance, the mobile host should have the ability to adapt

Table 1.2   Differences between Cellular and Ad Hoc Wireless Networks

Cellular	Ad Hoc Wireless Networks
Infrastructure networks	Infrastructureless networks
Fixed, prelocated cell sites and base station	No base station, and rapid deployment
Static backbone network topology	Highly dynamic network topologies with multihop
Relatively caring environment and stable connectivity	Hostile environment (noise, losses) and irregular connectivity
Detailed planning before base station can be installed	Ad hoc network automatically forms and adapts to changes
High setup costs	Cost-effective
Large setup time	Less setup time

to variation in performance and coverage and to switch gateways when beneficial. To enhance the prediction of the best overall performance, a network-layer metric has a better overview of the network. Ad hoc networking brings features like easy connection to access networks, dynamic multihop network structures, and direct peer-to-peer communication. The multihop property of an ad hoc network needs to be bridged by a gateway to the wired backbone. The gateway must have a network interface on both types of networks and be a part of both the global routing and the local ad hoc routing. Users could benefit from ubiquitous networks in several ways. User mobility enables users to switch between devices, migrate sessions, and still get the same personalized services. Host mobility enables the users' devices to move around the networks and maintain connectivity and reachability.

1.3.1 Differences between Cellular and Ad Hoc Wireless Networks

Table 1.2 gives the major differences between cellular and ad hoc networks.

1.3.2 Applications of Ad Hoc Wireless Networks

The field of wireless networking emerges from the integration of personal computing, cellular technology, and the Internet. This is due to the increasing interactions between communication and computing, which are changing information access from "anytime anywhere" into "all the time, everywhere." At present, a large variety of networks exists, ranging from the well-known infrastructure of cellular networks to noninfrastructure wireless ad hoc networks.

The following are the applications of ad hoc wireless networks:

■ Community network
■ Enterprise network
■ Home network
■ Emergency response network
■ Vehicle network
■ Sensor network

Unlike a fixed wireless network, wireless ad hoc or on-the-fly networks are characterized by the lack of infrastructure. Nodes in a mobile ad hoc network are free to move and organize themselves in an arbitrary fashion. Each user is free to roam about while communicating with others. The path between each pair of users may have multiple links, and the radio between them can be heterogeneous. This allows an association of various links to be a part of the same network. Mobile ad hoc networks can operate in a stand-alone fashion or could possibly be connected to a larger network such as the Internet.

Ad hoc networks are suited for use in situations where an infrastructure is unavailable or to deploy one is not cost-effective. One of many possible uses of mobile ad hoc networks is in some business environments, where the need for collaborative computing might be more important outside the office environment than inside, such as in a business meeting outside the office to brief clients on a given assignment. Work has been going on to introduce the fundamental concepts of game theory and its applications in telecommunications. Game theory originates from economics and has been applied in various fields. Game theory deals with multiperson decision making, in which each decision maker tries to maximize his or her utility. The cooperation of the users is necessary to the operation of ad hoc networks; therefore, game theory provides a good basis to analyze the networks.

A mobile ad hoc network can also be used to provide crisis management services applications, such as in disaster recovery, where the entire communication infrastructure is destroyed and resorting communication quickly is crucial. By using a mobile ad hoc network, an infrastructure could be set up in hours instead of weeks, as is required in the case of wired line communication. Another application example of a mobile ad hoc network is Bluetooth, which is designed to support a personal area network (PAN) by eliminating the need for wires between various devices, such as printers and personal digital assistants. The famous IEEE 802.11 or WiFi protocol also supports an ad hoc network system in the absence of a wireless access point.

The idea of ad hoc networking goes back to the U.S. Defense Advanced Research Projects Agency (DARPA) packet radio network, which was used in the 1970s. A mobile ad hoc network is a collection of mobile devices establishing a short live or temporary network in the absence of a supporting structure. Mobile ad hoc networks can be used in establishing efficient, dynamic communication for rescue, emergency, and military operations. A commercial application, such as Bluetooth, is one of the recent developments utilizing the concept of ad hoc networking.

Bluetooth is named after King Harald Blat (translated as King Harold Bluetooth in English), who ruled Denmark in the tenth century a.d. Bluetooth was first introduced in 1998. It uses radio waves to transmit wireless data over short distances, and can support many users in any environment. Eight devices can communicate with each other in a piconet. At one time, ten of these piconets can coexist in the same coverage range of the Bluetooth radio. A Bluetooth device can act as both a client and a server. A connection must be established to exchange data between any two Bluetooth devices. To establish a connection, a device must request a connection with the other device. Bluetooth was based on the idea of advancing wireless interactions with various electronic devices. Devices like mobile phones, personal digital assistants, and laptops with the right chips could all communicate wirelessly with each other. However, later it was realized that a lot more is possible.

1.3.3 Technical and Research Challenges

Mobile ad hoc networks pose several technical and research challenges that need to be addressed. Ad hoc architecture has many benefits, such as self-reconfiguration and adaptability to highly variable mobile characteristics such as power and transmission conditions, traffic distributions, and load balancing. These benefits pose new challenges which mainly reside in the unpredictability of network topology due to the mobility of nodes, which, coupled with the local broadcast capability, causes a set of concerns in designing a communication system on top of ad hoc wireless networks. To deal with this issue, many potential approaches have been proposed: distributed MAC and dynamic routing, Wireless Service Location Protocol, Wireless Dynamic Host Configuration Protocol, distributed admission call control, and quality-of-service (QoS)–based routing technique.

1.3.3.1 Security Issues and Challenges

Security has become a primary concern to provide protected communication between mobile nodes in a hostile environment. Unlike the wired line networks, the unique characteristics of mobile ad hoc networks pose a number of nontrivial challenges to security design, such as open peer-to-peer network architecture, shared wireless medium, stringent resource constraints, and highly dynamic network topology. These challenges clearly make a case for building multifence security solutions that achieve both broad protection and desirable network performance.

One of the fundamental vulnerabilities of MANETs comes from their open peer-to-peer architecture. Unlike wired networks that have dedicated routers, each mobile node in an ad hoc network may function as a router and forward packets for other nodes. The wireless channel is accessible to both legitimate network users and malicious attackers. As a result, there is no clear line of defense in MANETs from the security design perspective. The boundary that separates the inside network from the outside world becomes blurred. There is no well-defined place or infrastructure where we may deploy a single security solution. Moreover, portable devices, as well as the system security information they store, are vulnerable to compromises or physical capture, especially low-end devices with weak protection. Attackers may sneak into the network through these subverted nodes, which pose the weakest link and incur a domino effect of security breaches in the system.

The stringent resource constraints in MANETs constitute another nontrivial challenge to security design. The wireless channel is bandwidth constrained and shared among multiple networking entities. The computation capability of a mobile node is also constrained. For example, some low-end devices, such as PDAs, can hardly perform computation-intensive tasks like asymmetric cryptographic computation. Because mobile devices are typically powered by batteries, they may have very limited energy resources. The wireless medium and node mobility pose far more dynamics in MANETs compared to the wired line networks. The network topology is highly dynamic as nodes frequently join or leave the network, and roam in the network on their own will. The wireless channel is also subject to interferences and errors, exhibiting volatile characteristics in terms of bandwidth and delay. Despite such dynamics, mobile users may request "anytime, anywhere" security services as they move from one place to another.

The above characteristics of MANETs clearly make a case for building multi-fence security solutions that achieve both broad protection and desirable network performance. First, the security solution should spread across many individual components and rely on their collective protection power to secure the entire network. The security scheme adopted by each device has to work within its own resource limitations in terms of computation capability, memory, communication capacity, and energy supply. Second, the security solution should span different layers of the protocol stack, with each layer contributing to a line of defense. No single-layer solution is possible to thwart all potential attacks. Third, the security solution should thwart threats from both outsiders who launch attacks on the wireless channel and network topology, and insiders who sneak into the system through compromised devices and gain access to certain system knowledge. Fourth, the security solution should encompass all three components of prevention, detection, and reaction that work in concert to guard the system from collapse. Finally, the security solution should be practical and affordable in a highly dynamic and resource-constrained networking scenario.

1.3.3.2 Different Types of Attacks on Multicast Routing Protocols

Rushing attack: Many demand-driven protocols such as On-Demand Multicast Routing Protocol (ODMRP), Multicast Ad Hoc On-Demand Distance Vector

(MAODV), and Adaptive Demand-Driven Multicast Routing Protocol (ADMR), which use the duplicate suppression mechanism in their operations, are vulnerable to rushing attacks. When source nodes flood the network with route discovery packets to find routes to the destinations, each intermediate node processes only the first nonduplicate packet and discards any duplicate packets that arrive at a later time. Rushing attackers, by skipping some of the routing processes, can quickly forward these packets and be able to gain access to the forwarding group.

Black hole attack: First, a black hole attacker needs to invade into the forwarding group, for example by implementing a rushing attack, to route data packets for some destination to itself. Then, instead of doing the forwarding task, the attacker simply drops all of the data packets it receives. This type of attack often results in a very low packet delivery ratio.

Neighbor attack: Upon receiving a packet, an intermediate node records its ID in the packet before forwarding the packet to the next node. However, if an attacker simply forwards the packet without recording its ID in the packet, it makes two nodes that are not within the communication range of each other believe that they are neighbors (i.e., one hop away from each other), resulting in a disrupted route.

Jellyfish attack: Similar to the black hole attack, a jellyfish attacker first needs to intrude into the forwarding group and then it delays data packets unnecessarily for some amount of time before forwarding them. This result in significantly high end-to-end delay and delay jitter, and thus degrades the performance of real-time applications.

1.3.3.3 Interconnection of Mobile Ad Hoc Networks and the Internet

The interconnection of mobile ad hoc networks to fixed IP networks is one of the topics receiving more attention within the MANET working group of the IETF as well as in many research projects funded by the European Union. Several solutions have recently been proposed, but at this time it is unclear which ones offer the best performance compared to the others. In addition to introducing the main challenges and design, options that need to be considered are discussed in detail in this text.

1.3.4 Issues in Ad Hoc Wireless Networks

Different types of terminals form most of the ad hoc networks—for example, PDA-like devices, mobile phones, two-way pagers, sensors, or desktop computers—with different capabilities in terms of maximum transmission power, energy availability, mobility patterns, and QoS requirements. Ad hoc networks are generally heterogeneous in terms of terminals and services offered. In terms of energy and power, one has to consider not only node heterogeneity in terms of transmission power and energy availability, but also varying communication ranges, such as sleeping or active modes and the existence of energy supplies. Ad hoc networks raise new issues concerning security and privacy.

Ad hoc networks inherit some of the traditional problems of wireless communication and wireless networking:

- The wireless medium does not have proper boundaries outside of which nodes are known to be unable to receive network frames.
- The wireless channel is weak, unreliable, and unprotected from outside signals, which may cause lots of problems to the nodes in the network.
- The wireless channel has time-varying and asymmetric propagation properties.
- Hidden-node and exposed-node problems may occur.

1.3.4.1 Medium Access Control (MAC) Protocol Research Issues

Wireless multiple access can be categorized into random access (e.g., CSMA and CSMA with Collision Detection [CSMA/CD]) and controlled access (e.g., TDMA and token-based schemes). Random access will be suitable for ad hoc networks because of lack of infrastructure support. In addition, the IEEE 802.11 (WLAN) committee as the basis for its standards selected the CSMA/CA scheme. The Bluetooth technology that is designed to support, beyond data traffic and delay-sensitive applications (e.g., audio and video), adopted the TDMA scheme with an implicit token-passing scheme for the slots assignment. The use of Bluetooth and IEEE 802.11 is not optimized in a multihop environment. These technologies are used for single-hop Wireless Personal Area Networks (WPANs) and WLANs, respectively. The design of MAC protocols for a multihop ad hoc environment is a hot research issue.

1.3.4.2 Networking Issues

Most of the main functionalities of the *networking protocols* need to be redesigned. Networking protocols use one-hop transmission services provided by the enabling technologies to construct end-to-end delivery service from a sender needing to locate the receiver inside the network. The purpose of the *location services* is to dynamically map to its current location in the network. Current solutions generally adopted to manage mobile terminals in infrastructure networks are inadequate, and new approaches need to be found for mobile management.

A simple solution to node location is based on flooding the location query through the network. This approach is suitable for only limited-size networks. Controlling the flooding area can help to refine the technique. This can be achieved by gradually increasing, until the node is located, the number of hops involved in the flooding propagation.

The flooding approach constitutes a reactive location service in which no location information is maintained inside the network. The location information service maintenance cost is negligible, and all the complexity is associated with query operations. On the other hand, proactive location services subdivide the complexity into two phases. Proactive services construct and maintain inside the network data structures that store the location information of each node. By exploiting the data structures, the query operations are highly simplified.

1.3.4.3 Ad Hoc Routing and Forwarding

The highly dynamic nature of a mobile ad hoc network results in frequent and unpredictable changes of network topology, adding difficulty and complexity to routing among the mobile nodes. The challenges and complexities, coupled with the critical importance of routing protocol in establishing communications among mobile nodes, make the routing area the most active research area with the MANET domain.

Numerous routing protocols and algorithms have been proposed. Their performance under various network environments and traffic conditions has been studied and compared. Several surveys and comparative analyses of MANET routing protocols have been published. The classification of the routing protocols can be done via the type of cast property, that is, whether they use a Unicast, Multicast, or Geocast or Broadcast Routing Protocol.

1.3.4.4 Unicast Routing

A primary goal of Unicast Routing Protocols is the correct and efficient route establishment and maintenance between a pair of nodes, so that messages may be delivered reliably and in a timely manner. MANET characteristics make the direct use of these protocols infeasible. MANET Routing Protocols must operate in networks with highly dynamic topologies where routing algorithms run on resource-constrained devices.

MANET Routing Protocols are typically subdivided into two main categories: proactive routing protocols and reactive routing protocols. Proactive routing protocols are derived from distance-vector and link-state protocols. They maintain consistent and updated routing information for every pair of network nodes by propagating, proactively, route updates at fixed time intervals. As the routing information is usually maintained in tables, these protocols are also referred to as "Table-Driven protocols."

1.3.4.4.1 Proactive Routing Protocols

"Proactive routing protocol" is the constant maintaining of a route by each node to all other network nodes. The route creation and maintenance are performed through both periodic and event-driven messages. The various proactive protocols are Destination-Sequenced Distance-Vector (DSDV), Optimized Link State Routing (OLSR), and Topology Dissemination Based on Reverse Path Forwarding (TBRPF).

The DSDV protocol is a distance-vector protocol with extensions to make it suitable to MANET. Every node maintains a routing table with one route recorded. To avoid routing loops, a destination sequence number is used. A node increments its sequence number whenever a change occurs in its neighborhood.

The OLSR protocol is an optimization for MANET of legacy link-state protocols. The key point of the optimization is the MultiPoint Relay (MPR). By flooding a message to its MPRs, a node is guaranteed that the message, when retransmitted by the MPRs, will be received by all its two-hop neighbors. TBRPF is a link-state routing protocol that employs a different overhead reduction technique. Each node computes a shortest-path tree to all other nodes, but to optimize bandwidth only part of the tree is propagated to neighbors. The Fisheye State Routing (FSR) protocol is also an optimization over link-state algorithms using the fisheye technique. FSR propagates link-state information to other nodes in the network based on how far away the nodes are.

1.3.4.4.2 Reactive Routing Protocols

With these protocols, to reduce overhead, the route between two nodes is discovered only when it is needed. There are different types of reactive routing protocols such as Dynamic Source Routing (DSR), Ad Hoc On-Demand Distance Vector (AODV), Temporally Ordered Routing Algorithm (TORA), Associativity-Based Routing (ABR), and Signal Stability Routing (SSR).

DSR is a loop-free, source-based, on-demand routing protocol, where each node maintains a route cache that contains the source routes learned by the node. The route discovery process is initiated only when a source node does not already have a valid route to the destination in its route cache; entries in the route cache are continually updated as new routes are learned. Source routing is used for packet forwarding. AODV is another reactive improvement of the DSDV protocol. AODV minimizes the number of route broadcasts by creating routes on demand, as opposed to maintaining a complete list of routes as in the DSDV algorithm. Similar to DSR, route discovery is initiated on demand, and the route request is then forwarded by the source to the destination.

TORA is another source-initiated, on-demand routing protocol built on the concept of link reversal of the Directed Acyclic Graph (ACG). In addition to being loop free and bandwidth efficient, TORA has the properties of being highly adaptive and quick in route repair during link adaptivity and quick in route repair during link failure, while providing multiple routes for any desired source–destination pair.

The ABR protocol is also a loop-free protocol, built using a new routing metric termed "degree of association stability in selecting routes," so that routes discovered can be longer lived, and thus more stable and requiring less updates subsequently. The limitation of ABR comes mainly from a periodic, used to establish the association stability metrics, which may result in additional energy consumption. The Signal Stability Algorithm (SSA) is basically an ABR protocol with the additional property of route selection using the signal strength of the link.

1.3.4.4.3 Hybrid Protocols

In addition to proactive and reactive routing protocols, another class of Unicast Routing Protocols that can be identified is hybrid protocols. The Zone-Based Hierarchical Link-State Routing Protocol (ZRP) is an example of a hybrid protocol that combines both proactive and reactive approaches, thus trying to bring together the advantages of the two approaches. ZRP defines around each node a zone that contains the neighbors within a given number of hops from the node. Proactive and reactive algorithms are used by the node to route packets within and outside the zone, respectively.

1.3.4.5 *Multicast Routing*

Multicasting is an efficient communication service for supporting multipoint applications. Two main approaches are used for multicast routing in fixed networks: group-shared tree and source-specific tree. In the group share, a single tree is constructed for the whole group. The source-specific approach maintains, for each source, a tree toward all its receivers. There are two types of multicast protocols: MAODV and Ad Hoc Multicast Routing Protocol Utilizing Increasing ID-Numbers (AMRIS). Both protocols are on demand and construct a shared delivery tree to support multiple senders and receivers within a multicast session. The topology of a wireless mobile network can be very dynamic, and hence the maintenance of a connected multicast routing tree may cause large overhead. To avoid this, a different approach based on meshes has been proposed. Meshes are more suitable for dynamic environments because they support more connectivity than trees; thus they support multicast trees. There are two types of mesh-based multicast routing protocols: Core-Assisted Mesh Protocol (CAMP) and the On-Demand Multicast Routing Protocol (ODMRP). These protocols build routing meshes to disseminate multicast packets within groups. The difference is that

ODMRP uses flooding to build the mesh, whereas CAMP uses one or more nodes to assist in building the mesh, instead of flooding.

1.3.4.6 Location-Aware Routing

Location-aware routing protocols use, during forwarding operations, the node's position provided by Global Positioning System (GPS) or other mechanisms. Specifically, a node selects the next hop for packet forwarding by using the physical position of its one-hop neighbors and the physical position of the destination node. Location-aware routing does not require router establishment and maintenance. No routing information is stored. The use of geolocation information avoids networkwide searches, as both control and data packets are sent toward the known geographical coordinates of the destination node.

Three main strategies can be identified in location-aware routing protocols: greedy forwarding, directed flooding, and hierarchical routing.

> *Greedy forwarding*: In this type of strategy, a node tries to forward the packet to one of its neighbors that is closer to the destination than itself. If more than one closer node exists, different choices are possible. If, on the other hand, no closer neighbor exists, new rules are included in the greedy strategies to find an alternative route.
>
> *Direct flooding*: Directing flooding nodes forward the packets to all neighbors that are located in the direction of the destination. Distance Routing Effect Algorithm for Mobility (DREAM) and Location Aid Routing (LAR) are two routing algorithms that apply this principle.
>
> *Hierarchical routing*: The Location Proxy Routing Protocol and the Terminode Routing Protocol are hierarchical routing protocols in which routing is structured in two layers. Both protocols apply different rules to long- and short-distance routing, respectively. Location-aware routing is used for routing on long distances, whereas when a packet arrives close to the destination, a proactive distance-vector scheme is adopted.

1.3.4.7 Transmission Control Protocol (TCP) Issues

TCP is an effective connection-oriented transport control protocol that provides the essential flow control and congestion control required to ensure reliable packet delivery. Numerous enhancements and optimizations have been proposed over the past few years to improve TCP performance for infrastructure-based WLANs and cellular networking environments. Infrastructure-based wireless networks are one-hop wireless networks where a mobile device uses the wireless medium to access the fixed infrastructure. The mobile multihop ad hoc environment brings fresh challenges to TCP protocol.

The main research areas and open issues include the following:

- Impact of mobility
- Nodes interaction MAC layer
- Impact of TCP congestion window size
- Interaction between MAC protocols

1.3.4.8 Network Security

The wireless ad hoc nature of MANET brings new security challenges to the network design. Wireless networks are generally more vulnerable to information and physical security threats than fixed wired networks. Vulnerability of channels and nodes, absence of infrastructure, and dynamically changing topology make ad hoc networks' security a difficult task. Broadcast wireless channels allow message eavesdropping and injection. The absence of infrastructure makes the classic security solutions based on certification authorities and online servers inapplicable. Routing the packets in a secured environment is another challenge.

1.3.4.9 Different Security Attacks

Securing wireless ad hoc networks is a highly challenging issue. There are certain specific attacks to which the ad hoc context is vulnerable. Performing communication in free space exposes ad hoc networks to eavesdrop or inject messages. Ad hoc network attacks can be classified into active and passive attacks. A passive attack does not inject any message, but listens to the channel. A passive attack tries to discover valuable information and does not produce any new traffic in the network. In the case of an active attack, messages are inserted into the network; such attacks involve actions such as replication, modification, and deletion of exchanged data. In ad hoc networks, active attacks are impersonation, Denial Of Service (DOS), and disclosure attack.

Impersonation: In this type of attack, nodes may join the network undetectably, or send false routing information, masquerading as some other trusted node. A black hole attack falls in this category: here, a malicious node uses the routing protocol to advertise itself as having the shortest path to the node whose packets it wants to intercept.
Denial of service: Attacks like routing table overflow and sleep deprivation fall in this category.
Disclosure attack: A location disclosure attack can reveal something about the physical location of nodes or the structure of the network. Two types of security mechanisms can be generally applied: preventive and detective. Preventive mechanisms are typically based on key-based cryptography. Key distribution is at the

center of prevent mechanisms, since no central authority, no centralized trusted third party, and no central server are available ad hoc. Detective mechanisms have to monitor and rely on the audit trace that is limited to communication activities taking place within the radio range.

1.3.4.10   Security at Data-Link Layer

The wireless medium access protocol implements mechanisms based on cryptography to avoid unauthorized access and to enhance the privacy on radio links. An analysis of IEEE 802.11 and Bluetooth can be discussed in brief.

Security in the IEEE 802.11 standard is provided by the Wired Equivalent Privacy (WEP) scheme, which supports both data encryption and integrity. The key is a 40-bit secret key and is shared only by all the devices of a WLAN, or is a pairwise secret key shared only by two communicating devices.

Bluetooth uses cryptographic security mechanisms implemented in the data-link layer. A key management service provides each device with a set of symmetric cryptographic keys required for the initialization of a secret channel with another device, the execution of an authentication protocol, and the exchange of encrypted data on the secret channel.

1.3.4.11   Secure Routing

Malicious nodes can disrupt the correct functioning of a routing protocol by *modifying* routing information, *fabricating* false routing information, and *impersonating* other nodes. The Secure Routing Protocol (SRP) is an extension that is applied to several existing reactive routing protocols. SRP is based on the assumption of the existence of a security association between the sender and receiver based on a shared secret key negotiated at the connection setup. SRP fights against the attacks that disrupt the route discovery process. A node initiating a route discovery is able to identify and discard false routing information. Ariadne is a secure ad hoc routing protocol based on DSR and the Timed Efficient Stream Loss-Tolerant Authentication (TESLA) authentication protocol.

The Authenticated Routing for Ad Hoc Network (ARAN) Protocol is on-demand, secure, and detects and protects against malicious actions carried out by third parties in the ad hoc environment. ARAN is based on certificates from a trusted certificate server before joining the ad hoc network.

Secure Efficient Ad Hoc Distance (SEAD) is a proactive secure routing protocol based on routing table update messages. The basic idea is to authenticate the sequence number and the metric field of a routing table update message using one-way hash functions. Hash chains and digital signatures are used by the Secure Ad Hoc On-Demand Distance Vector (SAODV) mechanism.

1.3.4.12 Quality of Service (QoS)

The ability of a network to provide QoS depends on the intrinsic characteristics of all the network components, from transmission links to MAC and network layers. Wireless links have a low and highly variable capacity, and high loss rates. Topologies are highly dynamic and have high packet loss rates. Random access-based MAC protocols have no QoS support.

QoS MAC protocols solve the problems of medium contention, support reliable unicast communications, and provide resource reservation for real-time traffic in a distributed wireless environment. Numerous MAC protocols and improvements that have proposed protocols that can provide QoS guarantees to real-time traffic in a distributed wireless environment include Group Allocation Multiple Access with Piggyback Reservation (GAMA/PR) protocol and Black Burst (BB) contention mechanism.

1.4 Problems

1.1   Give the features of IrDA with suitable illustrations.
1.2   Compare and contrast Bluetooth with IrDA.
1.3   List the features of HomeRF.
1.4   Explain a typical Wireless Network with a suitable illustration.
1.5   Describe the advantages and disadvantages of WiFi and WiMax.
1.6   Discuss the technology deployed in wireless Internet.
1.7   Give the limitations of IP.
1.8   Discuss the working of datagram routing using mobile IP.
1.9   Explain the main issues involved in mobile IP.
1.10 What are Ad hoc Networks? Explain.
1.11 Differentiate between cellular and ad hoc wireless networks.
1.12 Give the applications of ad hoc networks.
1.13 Discuss in detail the technical and research challenges in ad hoc networks.
1.14 Describe the issues in ad hoc networks.
1.15 Explain the security problems in ad hoc networks.

Bibliography

C. Barrett et al., Characterizing the interaction between routing and MAC protocols in ad-hoc networks, Proc. MobiHoc, 2002, pp. 92–103.

J. Broch et al., A performance comparison of multi-hop wireless ad hoc network routing protocols, Proc. Mobicom, 1998.

M. Frodigh et al., Wireless ad hoc networking: The art of networking without a network, *Ericsson Review*, No. 4, 2000.

Z. J. Haas et al., (Eds.), Special Issue on Wireless ad hoc networks, *IEEE Journal on Selected Areas in Communications*, 17(8), Aug. 1999.

T. Larsson and N. Hedman, Routing protocols in wireless ad-hoc networks—A simulation study, Master's thesis at Lulea University of Technology, Stockholm, 1998.

R. Ramanathan, Making ad hoc networks density adaptive, Proc. Milcom, 2001, pp. 957–961.

S. Ramanathan and M. Steenstrup, A survey of routing techniques for mobile communications networks, Baltzer/ACM *Mobile Networks and Applications*, No. 1, 1996, pp. 89–104.

V. Ramasubramanian et al., SHARP: A hybrid adaptive routing protocol for mobile ad hoc networks, Proceedings of 4th International Symposium on Mobile Ad Hoc Networking and Computing (MobiHoc), Annapolis, MD, June 3–6, 2003.

N. Roux et al., Cost adaptive mechanism to provide network diversity for MANET reactive routing protocols, Proceedings MILCOM, 2000.

E. M. Royer and C.-K. Toh, A review of current routing protocols for ad hoc mobile wireless networks, *IEEE Personal Communications*, April 1999, pp. 46–55.

Chapter 2

MAC Layer Protocols for Ad Hoc Wireless Networks

2.1 Introduction

The simplicity in deployment makes mobile ad hoc networks (which do not require any infrastructure) suitable for a variety of applications such as collaborative computing, disaster recovery, and battlefield communication. With the proliferation of communications and computing devices such as mobile phones, laptops, or Personal Digital Assistants (PDAs), Personal Area Networking (PAN), which is an ad hoc networking-based technology, has recently gained much interest.

In ad hoc networks, transmitters use radio signals for communication. Generally, each node can only be a transmitter (TX) or a receiver (RX) at a time. Communication among mobile nodes is limited within a certain transmission range. And nodes share the same frequency domain to communicate. So, within such a range, only one transmission channel is used, covering the entire bandwidth. Unlike wired networks, packet delay is caused by not only the traffic load at the node, but also the traffic load at the neighboring nodes, which is called "traffic interference."

Medium Access Control (MAC) protocols play an important role in the performance of the Mobile Ad hoc Networks (MANETs). A MAC protocol defines how each mobile unit can share the limited wireless bandwidth resource in an efficient manner. The source and destination could be far away, and each time packets need to be relayed from one node to another in multihop fashion, a medium has to be accessed. Accessing a medium properly requires only informing the nodes within the vicinity of transmission. MAC protocols control access to the transmission medium.

Their aim is to provide an orderly and efficient use of the common spectrum. These protocols are responsible for per-link connection establishment (i.e., acquiring the medium) and per-link connection cancellation (i.e., releasing the medium).

One of the fundamental challenges in MANET research is how to increase the overall network throughput while maintaining low energy consumption for packet processing and communications. The low throughput is attributed to the harsh characteristics of the radio channel combined with the contention-based nature of MAC protocols commonly used in MANETs.

Regarding the MAC protocol for a wireless MANET, the following performance measures should be considered:

- *Throughput and delay*: Throughput is generally measured as the percentage of successfully transmitted radio-link level frames per unit time. Transmission delay is defined as the interval between the frame arrival time at the MAC layer of a transmitter and the time at which the transmitter realizes that the transmitted frame has been successfully received by the receiver.
- *Fairness*: Generally, fairness measures how fair the channel allocation is among the flows in the different mobile nodes. The node mobility and the unreliability of radio channels are the two main factors that impact fairness.
- *Energy efficiency*: Generally, energy efficiency is measured as the fraction of the useful energy consumption (for successful frame transmission) to the total energy spent.
- *Multimedia support*: This is the ability of a MAC protocol to accommodate traffic with different service requirements, such as throughput, delay, and frame loss rate.

2.2 Important Issues and the Need for Medium Access Control (MAC) Protocols

There are several important issues in ad hoc wireless networks. Most ad hoc wireless network applications use the Industrial, Scientific and Medical (ISM) band that is free from licensing formalities. Because wireless is a tightly controlled medium, it has a limited channel bandwidth that is typically much smaller than that of wired networks. Besides, the wireless medium is inherently error prone. Even though a radio may have sufficient channel bandwidth, factors such as multiple access, signal fading, and noise and interference can cause the effective throughput in wireless networks to be significantly lower. Because wireless nodes may be mobile, the network topology can change frequently without any predictable pattern. Usually the links between nodes would be bidirectional, but there may be cases when differences in transmission power give rise to unidirectional links, which necessitate special treatment by the MAC protocols. Ad hoc network nodes must conserve energy as they mostly rely on batteries as their power source. Security issues should be considered in the overall network design, as it is relatively easy to eavesdrop on wireless transmission. Routing protocols require information about the current topology, so that a route from a source to a destination may be found. However, the existing routing schemes, such as distance-vector and link-state-based protocols, lead to poor route convergence and low throughput for dynamic topology. Therefore, a new set of routing schemes is needed in the ad hoc wireless context. The MAC layer, sometimes also referred to as a sublayer of the data-link layer, involves the functions and procedures necessary to transfer data between two or more nodes of the network. It is the responsibility of the MAC layer to perform error correction for anomalies occurring in the physical layer. The layer performs specific activities for framing, physical addressing, and flow and error controls. It is responsible for resolving conflicts among different nodes for channel access. Because the MAC layer has a direct bearing on how reliably and efficiently data can be transmitted between two nodes along the routing path in the network, it affects the quality of service (QoS) of the network. The design of a MAC protocol should also address issues caused by mobility of nodes and an unreliable time-varying channel.

2.2.1 Need for Special MAC Protocols

The popular Carrier Sense Multiple Access (CSMA) MAC scheme and its variations such as CSMA with Collision Detection (CSMA/CD), developed for wired networks, cannot be used directly in the wireless networks, as will be explained below. In CSMA-based schemes, the transmitting node first senses the medium to check whether it is idle or busy. If the medium is busy, the node defers its own transmission to prevent a collision with the existing signal. Otherwise, the node begins to transmit its data while continuing to sense the medium. However, collisions occur at receiving nodes. Because signal strength in the wireless medium fades in proportion to the square of distance from the transmitter, the presence of a signal at the receiver node may not be clearly detected at other sending terminals, if they are out of range. As illustrated in Figure 2.1, node B is within the range of nodes A and C, but A and C are not in each other's range. Let us consider the case where A is transmitting to B. Node C, being out of A's range, cannot detect a carrier and may therefore send data to B, thus causing a collision at B. This is referred to as the "hidden-terminal problem," as nodes A and C are hidden from each other. Let us now consider another case where B is transmitting to A. Because C is within B's range, it senses a carrier and decides to defer its own transmission. However, this is unnecessary because there is no way C's transmission can cause any collision at receiver A. This is referred to as the "exposed-terminal problem," because B being exposed to C caused the latter to needlessly defer its transmission. MAC schemes are designed to overcome these problems.

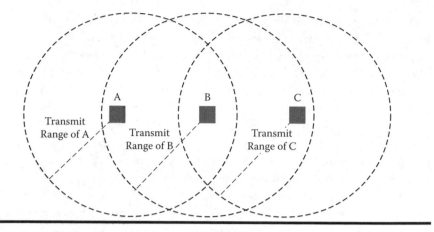

Figure 2.1   Illustration of hidden- and exposed-terminal problem.

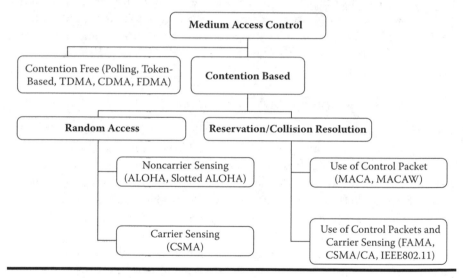

Figure 2.2   Classification of MAC protocols.

2.3 Classification of MAC Protocols

This section describes the classification of MAC protocols and the various factors considered for classification. Various MAC schemes developed for wireless ad hoc networks can be classified as shown in Figure 2.2. In contention-free schemes (e.g., Time Division Multiple Access [TDMA], Frequency Division Multiple Access [FDMA], and Code Division Multiple Access [CDMA]), certain assignments are used to avoid contentions. Contention-based schemes, on the other hand, are aware of the risk of collisions of transmitted data. Because contention-free MAC schemes are more applicable to static networks or networks with centralized control, this chapter focuses on contention-based MAC schemes.

Another category depends on energy-efficient protocols at all layers of the network model. Hence, MAC protocols that are power aware are needed. Another class of MAC protocols uses directional antennas. The advantage of this method is that the signals are transmitted in only one direction. The nodes in other directions are, therefore, no longer prone to interference or collision effects, and spatial reuse is facilitated. Several MAC schemes have been proposed for unidirectional links.

Users will demand some level of QoS from a MANET regarding such issues as end-to-end delay, available bandwidth, and probability of packet loss. However, the lack of centralized control, limited bandwidth channels, node mobility, and power or computational constraints, and the error-prone nature of the wireless medium, make it very difficult to provide effective QoS in ad hoc networks. Because the MAC layer has a direct bearing on how reliably and efficiently data can be transmitted from one node to the next along the routing path in the network, it affects the QoS of the network. Several QoS-aware MAC schemes are discussed in this chapter.

Another classification is based on the number of channels used for data transmission. Single-channel protocols set up reservations for transmissions, and subsequently transmit their data using the same channel or frequency. Many MAC schemes use a single channel. Multiple-channel protocols, as their name implies, use more than one channel to coordinate connection sessions among the transmitter and receiver nodes.

2.3.1 Contention-Based MAC Protocols

These protocols concentrate on the collisions of transmitted data. This includes two categories, Random Access Protocols and Dynamic Reservation–Collision Resolution Protocols.

1. With random access-based schemes, such as ALOHA, a node may access the channel as soon as it is ready. Naturally, more than one node may transmit at the same time, causing collisions. ALOHA is more suitable under low system loads with a large number of potential senders, and it offers relatively low throughput. A variation of ALOHA, Slotted ALOHA, introduces synchronized transmission time slots similar to TDMA. In this case, nodes can transmit only at the beginning of a time slot. The introduction of time slots doubles the throughput as compared to the pure ALOHA scheme, with the cost of necessary time synchronization. The CSMA-based schemes further reduce the possibility of packet collisions and improve the throughput.

2. Dynamic Reservation–Collision Resolution Protocols: To solve the hidden- and exposed-terminal problems in CSMA, researchers have come up with many protocols that are contention based but involve some forms of dynamic reservation and collision resolution. Some schemes use the request-to-send/ clear-to-send (RTS/CTS) control packets to prevent collisions, for example

multiple access collision avoidance (MACA) and MACA for wireless LANs (MACAW).

The contention-based MAC schemes can also be classified as sender-initiated versus receiver-initiated, single-channel versus multiple-channel, power-aware, directional antenna–based, unidirectional link–based, and QoS-aware schemes, as mentioned above. One distinguishing factor for MAC protocols is whether they rely on the sender initiating the data transfer, or the receiver requesting the same.

2.3.2 Contention-Based MAC Protocols with Reservation Mechanisms

The dynamic reservation approach involves setting up some sort of a reservation prior to data transmission. If a node that wants to send data takes the initiative of setting up this reservation, the protocol is considered to be a sender-initiated protocol. Most schemes are sender initiated. In a receiver-initiated protocol, the receiving node polls a potential transmitting node for data. If the sending node indeed has some data for the receiver, it is allowed to transmit it after being polled. The MACA—by Invitation (MACA-BI) and Receiver-Initiated—Busy Tone Multiple Access (RI-BTMA) are examples of such schemes.

2.3.2.1 Multiple Access Collision Avoidance (MACA)

The MACA protocol overcomes the hidden- and exposed-terminal problems. MACA uses two short signaling packets. The key idea of the MACA scheme is that any neighboring node that overhears an RTS packet has to defer its own transmissions until some time after the associated CTS packet would have finished, and that any node overhearing a CTS packet would defer for the length of the expected data transmission. In a hidden-terminal scenario, as explained in Section 2.2 (above), C will not hear the RTS sent by A, but it would hear the CTS sent by B. Accordingly, C will defer its transmission during A's data transmission. Similarly, in the exposed-terminal situation, C would hear the RTS sent by B, but not the CTS sent by A. Therefore C will consider itself free to transmit during B's transmission. It is apparent that this RTS–CTS exchange enables nearby nodes to reduce the collisions at the receiver, not the sender. Collisions can still occur between different RTS packets, though. If two RTS packets collide for any reason, each sending node waits for a randomly chosen interval before trying again. This process continues until one of the RTS transmissions elicits the desired CTS from the receiver. MACA is effective because RTS and CTS packets are significantly shorter than the actual data packets, and therefore collisions among them are less expensive compared to collisions among the longer data packets. However, the RTS–CTS approach does not always solve the hidden-terminal problem completely, and

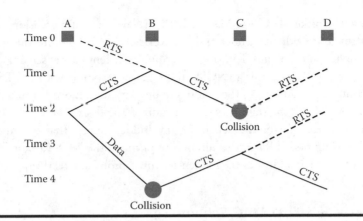

Figure 2.3 Illustration of failure of RTS–CTS mechanism in solving hidden- and exposed-terminal problems.

collisions can occur when different nodes send the RTS and the CTS packets. Let us consider an example with four nodes—A, B, C, and D—in Figure 2.3. Node A sends an RTS packet to B, and B sends a CTS packet back to A. At C, however, this CTS packet collides with an RTS packet sent by D. Therefore, C has no knowledge of the subsequent data transmission from A to B. While the data packet is being transmitted, D sends out another RTS because it did not receive a CTS packet in its first attempt. This time, C replies to D with a CTS packet that collides with the data packet at B. In fact, when hidden terminals are present and the network traffic is high, the performance of MACA degenerates to that of ALOHA. Another weakness of MACA is that it does not provide any acknowledgment of data transmissions at the data-link layer. If a transmission fails for any reason, retransmission has to be initiated by the transport layer. This can cause significant delays in the transmission of data.

2.3.2.1.1 MACA for Wireless LANs (MACAW)

This scheme uses a five-step (RTS–CTS–DS–DATA–ACK) exchange. MACAW allows much faster error recovery at the data-link layer by using the acknowledgment packet (ACK) that is returned from the receiving node to the sending node as soon as data reception is completed. The backoff and fairness issues among active nodes were also investigated. MACAW achieves significantly higher throughput compared to MACA. It does not, however, fully solve the hidden- and exposed-terminal problems.

2.3.2.1.2 Floor Acquisition Multiple Access (FAMA)

FAMA is another MACA-based scheme that requires every transmitting station to acquire control of the floor (i.e., the wireless channel) before it actually sends any data packet. Unlike MACA or MACAW, FAMA requires that collision avoidance be performed at both the sender and receiver nodes. To "acquire the floor," the sending node sends out an RTS using either nonpersistent packet sensing (NPS) or Nonpersistent Carrier Sensing (NCS). The receiver responds with a CTS packet, which contains the address of the sending node. Any station overhearing this CTS packet knows about the station that has acquired the floor. The CTS packets are repeated long enough for the benefit of any hidden sender that did not register another sending node's RTS. The authors recommend the NCS variant for ad hoc networks because it addresses the hidden-terminal problem effectively.

2.3.2.2 IEEE 802.11 MAC Scheme

The IEEE 802.11 standard specifies two modes of MAC protocol: Distributed Coordination Function (DCF) mode (for ad hoc networks) and Point Coordination Function (PCF) mode (for centrally coordinated infrastructure-based networks). The DCF in IEEE 802.11 is based on CSMA with Collision Avoidance (CSMA/CA) protocol, which can be seen as a combination of the CSMA and MACA schemes. The protocol uses the RTS–CTS–DATA–ACK sequence for data transmission. Not only does the protocol use physical carrier sensing, but it also introduces the novel concept of virtual carrier sensing. This is implemented in the form of a Network Allocation Vector (NAV), which is maintained by every node. The NAV contains a time value that represents the duration up to which the wireless medium is expected to be busy because of transmissions by other nodes. Because every packet contains the duration information for the remainder of the message, every node overhearing a packet continuously updates its own NAV. Time slots are divided into multiple frames, and there are several types of InterFrame Spacing (IFS) slots. In increasing order of length, they are the Short IFS (SIFS), Point Coordination Function IFS (PIFS), DCF IFS (DIFS), and Extended IFS (EIFS). The node waits for the medium to be free for a combination of these different times before it actually transmits. Different types of packets can require the medium to be free for a different number or type of IFS. For instance, in ad hoc mode, if the medium is free after a node has waited for DIFS, it can transmit a queued packet. Otherwise, if the medium is still busy, a backoff timer is initiated. The initial backoff "10" value of the timer is chosen randomly from between 0 and CW-1, where CW is the width of the contention window, in terms of time slots. After an unsuccessful transmission attempt, another backoff is performed with a doubled size of CW as decided by a Binary Exponential Backoff (BEB) algorithm. Each time the medium is idle after DIFS, the timer is decremented. When the timer expires, the packet is transmitted. After each successful transmission, another random backoff (known as "postback-off") is performed by the transmission-completing node. A control packet such as RTS, CTS, or ACK is transmitted after the medium has been free for SIFS.

2.3.2.3 Multiple Access Collision Avoidance—
by Invitation (MACA-BI)

In typical sender-initiated protocols, the sending node needs to switch to receive mode (to get CTS) immediately after transmitting the RTS. Each such exchange of control packets adds to turnaround time, reducing the overall throughput. MACA-BI is a receiver-initiated protocol, and it reduces the number of such control packet exchanges. Instead of a sender waiting to gain access to the channel, MACA-BI requires a receiver to request the sender to send the data, by using a "Ready-To-Receive" (RTR) packet instead of the RTS and the CTS packets. Therefore, it is a two-way exchange (RTR–DATA) as against the three-way exchange (RTS–CTS–DATA) of MACA. Because the transmitter cannot send any data before being asked by the receiver, there has to be a traffic prediction algorithm built into the receiver so it can know when to request data from the sender. The efficiency of this algorithm determines the communication throughput of the system. The algorithm proposed by the authors piggybacks the information regarding packet queue length and data arrival rate at the sender in the data packet. When the receiver receives this data, it is able to predict the backlog in the transmitter and send further RTR packets accordingly. There is a provision for a transmitter to send an RTS packet if its input buffer overflows. In such a case, the system reverts to MACA. The MACA-BI scheme works efficiently in networks with predictable traffic patterns. However, if the traffic is bursty, the performance degrades to that of MACA.

2.3.2.4 Group Allocation Multiple Access with
Packet Sensing (GAMA-PS)

GAMA-PS incorporates features of contention-based as well as contention-free methods. It divides the wireless channel into a series of cycles. Every cycle is divided into two parts for contention and group transmission. Although the group transmission period is further divided into individual transmission periods, GAMA-PS does not require clock or time synchronization among different member nodes. Nodes wishing to make a reservation for access to the channel employ the RTS–CTS exchange. However, a node will back off only if it understands an entire packet. Carrier sensing alone is not a sufficient reason for backing off. GAMA-PS organizes nodes into transmission groups, which consist of nodes that have been allocated a transmission period. Every node in the group is expected to listen in on the channel. Therefore, there is no need of any centralized control. Every node in the group is aware of all the successful RTS–CTS exchanges and, by extension, of any idle transmission periods. Members of the transmission group take turns transmitting data, and every node is expected to send a Begin Transmission Period (BTP) packet before sending actual data. The BTP contains the state of the transmission group, the position of the node within that group, and the number of group members. A member station can transmit up to a fixed length of data, thereby increasing efficiency. The last member of the transmission group broadcasts a Transmit Request (TR) packet after it sends its data. Use of the TR shortens the maximum length of the contention period by forcing any station that might contend for group membership to do so at the start of the contention period. GAMA-PS assumes that there are no hidden terminals. As a result, this scheme may not work well for mobile ad hoc networks. When there is not enough traffic in the network, GAMA-PS behaves almost like CSMA. However, as the load grows, it starts to mimic TDMA and allows every node to transmit once in every cycle.

2.3.3 MAC Protocols Using Directional Antennas

MAC protocols for ad hoc networks typically assume the use of omnidirectional antennas, which transmit radio signals to and receive them from all directions. These MAC protocols require all other nodes in the vicinity to remain silent. With directional antennas, it is possible to achieve higher gain and restrict the transmission to a particular direction. Similarly, packet reception at a node with a directional antenna is not affected by interference from other directions. As a result, it is possible that two pairs of nodes located in each other's vicinity communicate simultaneously, depending on the direction of transmission. This would lead to better spatial reuse in the other, unaffected directions. Using these antennas, however, is not a trivial task as the correct direction should be provided and turned to in real time. Besides, new protocols would need to be designed for taking advantage of the new features enabled by directional antennas because the current protocols (e.g., IEEE 802.11) cannot benefit from these features. Currently, directional antenna hardware is considerably bulkier and more expensive than omnidirectional antennas of comparable capabilities. Applications involving large military vehicles are, however, suitable candidates for wireless devices using such antenna systems. The use of higher frequency bands (e.g., ultrawide band transmission) will reduce the size of directional antennas. Many schemes are proposed with this idea:

1. A Slotted ALOHA scheme with packet radio networks and directional antenna packet radio networks involving multiple and directional antennas. Channel-access models link power control and directional neighbor discovery in the context of beam-forming directional antennas.
2. A scheme in which every node dynamically stores some information about its neighbors and their transmission schedules through the use of special control packets. This allows a node to steer its antenna appropriately based on the ongoing transmissions in the neighborhood. A method for using the directional antennas to implement a new form of link-state-based routing is also proposed.
3. A Directional MAC (D-MAC) scheme using directional antennas uses the familiar RTS–CTS–DATA–ACK sequence where only the RTS packet is sent using a directional antenna. Every node is assumed to be equipped with

several directional antennas, but only one of them is allowed to transmit at any given time, depending on the location of the intended receiver. In this scheme, every node is assumed to be aware of its own location as well as the locations of its immediate neighbors. This scheme gives better throughput than IEEE 802.11 by allowing simultaneous transmissions that are not possible in current MAC schemes.

4. Based on the IEEE 802.11 protocol, a simple scheme was proposed in which every node has multiple antennas. Any node that has data to send first sends out an RTS in all directions using every antenna. The intended receiver also sends out the CTS packet in all directions using all the antennas. The original sender is now able to discern which antenna picked up the strongest CTS signal and can learn the relative direction of the receiver. The data packet is sent using the corresponding directional antenna in the direction of the intended receiver. Thus, the participating nodes need not know their location information in advance. Please note that only one radio transceiver in a node can transmit and receive at a time. This scheme can achieve up to 2 to 3 times better average throughput than CSMA/CA with the RTS–CTS scheme (using omnidirectional antennas).

5. A Multihop RTS MAC (M-MAC) scheme for transmission on multihop paths. Because directional antennas have a higher gain and transmission range than omnidirectional antennas, it is possible for a node to communicate directly with another node that is far away. M-MAC therefore uses multiple hops to send RTS packets to establish links between distant nodes, but the subsequent CTS, DATA, and ACK packets are sent in a single hop. This protocol can achieve better throughput and end-to-end delay than the basic IEEE 802.11 and D-MAC schemes (above). The performance also depends on the topology configuration and flow patterns in the system. The use of directional antennas can introduce three new problems: new kinds of hidden terminals, higher directional interference, and deafness. These problems depend on the topology and flow patterns. For example, the deafness is a problem if routes of two flows share a common link. Similarly, nodes that are in a straight line witness higher directional interference. The performance of these schemes will degrade with node mobility. Some of the current protocols inaccurately assume that the gain of a directional antenna is the same as that of an omnidirectional antenna. Similarly, none of them considers the effect of transmit power control, the use of multiple channels, and support for real-time traffic.

2.3.4 Multiple-Channel MAC Protocols

A major problem of single shared-channel schemes is that the probability of collision increases with the number of nodes. It is possible to solve this problem with multichannel approaches. As seen in the classification, some multichannel schemes use a dedicated channel for control packets (or signaling) and one separate channel for data transmissions. They set up busy tones on the control channel, one with small bandwidth consumption, so that nodes are aware of ongoing transmissions. Another approach is to use multiple channels for data packet transmissions. This approach has the following advantages:

1. Because the maximum throughput of a single-channel scheme is limited by the bandwidth of that channel, using more channels appropriately can potentially increase the throughput.
2. Data transmitted on different channels does not interfere with each other, and multiple transmissions can take place in the same region simultaneously. This leads to significantly fewer collisions.
3. It is easier to support QoS by using multiple channels.

In general, a multiple-data-channel MAC protocol has to assign different channels to different nodes in real time. The issue of medium access still needs to be resolved. This involves deciding, for instance, the time slots at which a node would get access to a particular channel. In certain cases, it may be necessary for all the nodes to be synchronized with each other, whereas in other instances, it may be possible for the nodes to negotiate schedules among themselves. The details of some of the multiple-channel MAC schemes are discussed below.

2.3.4.1 Dual Busy Tone Multiple Access (DBTMA)

In the schemes based on the exchange of RTS–CTS dialogue, these control packets themselves are prone to collisions. Thus, in the presence of hidden terminals, there remains a risk of subsequent data packets being destroyed because of collisions. The DBTMA scheme uses out-of-band signaling to effectively solve the hidden- and the exposed-terminal problems. Data transmission is, however, on the single shared wireless channel. It builds upon earlier work on the Busy Tone Multiple Access (BTMA) and the RI-BTMA schemes. DBTMA decentralizes the responsibility of managing access to the common medium and does not require time synchronization among the nodes. As in several schemes discussed earlier, DBMTA sends RTS packets on data channels to set up transmission requests. Subsequently, two different busy tones on a separate narrow channel are used to protect the transfer of the RTS and data packets. The sender of the RTS sets up a Transmit-Busy Tone (BTt). Correspondingly, the receiver sets up a Receive-Busy Tone (BTr) to acknowledge the RTS, without using any CTS packet. Any node that senses an existing BTr or BTt defers from sending its own RTS over the channel. Therefore, both of these busy tones together guarantee protection from collision from other nodes in the vicinity. Through the use of the BTt and BTr in conjunction, exposed terminals are able to initiate data packet transmissions. Also, hidden terminals can reply to RTS requests as simultaneous data transmission occurs between the receiver and sender. However, the DBTMA scheme does not use ACK to acknowledge the received data packets. It also requires additional hardware complexity.

2.3.4.2 Multichannel Carrier Sense Multiple Access (CSMA) MAC Protocol

The multichannel CSMA protocol divides the total available bandwidth (W) into N distinct channels of W/N bandwidth each. Here, N may be lower than the number of nodes in the network. Also, the channels may be divided based on either an FDMA or a CDMA. A transmitter would use carrier sensing to see if the channel it last used is free or not. It uses the last used channel if it is found free. Otherwise, another free channel is chosen at random. If no free channel is found, the node should back off and retry later. Even when the traffic load is high and sufficient channels are not available, chances of collisions are somewhat reduced because each node tends to prefer its last-used channel instead of simply choosing a new channel at random. This protocol has been shown to be more efficient than single-channel CSMA schemes. Interestingly, the performance of this scheme is lower than that of the single-channel CSMA scheme at lower traffic load or when there are only a small number of active nodes for a long period of time. This is due to the waste of idling channels. The protocol is extended to select the best channel based on the signal power observed at the sender side.

2.3.4.3 Hop-Reservation Multiple Access (HRMA)

HRMA is an efficient MAC protocol based on FHSS radios in the ISM band. It uses time-slotting properties of very slow FHSS such that an entire packet is sent in the same hop. HRMA requires no carrier sensing, employs a common frequency-hopping sequence, and allows a pair of nodes to reserve a frequency hop (through the use of an RTS–CTS exchange) for communication without interference from other nodes. One of the N available frequencies in the network is reserved specifically for synchronization. The remaining $N-1$ frequencies are divided into M = floor $((N-1)/2)$ pairs of frequencies.

For each pair, the first frequency is used for Hop Reservation (HR), RTS, CTS, and data packets, whereas the second frequency is used for ACK packets. HRMA can be treated as a TDMA scheme, where each time slot is assigned a specific frequency and subdivided into four parts—Synchronizing (SYN), HR, RTS, and CTS periods. Figure 2.4 shows an example of the HRMA frame.

During the synchronization period of every time slot, all idle nodes synchronize to each other. On the other three periods, they hop together on the common frequency hops that have been assigned to the time slots. A sender node first sends an RTS packet to the receiver in the RTS period of the time slot. The receiver

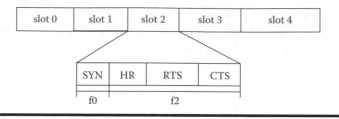

Figure 2.4   HRMA frame.

sends a CTS packet to the sender in the CTS period of that same time slot. Now, the sender sends the data on the same frequency (at this time, the other idle nodes are synchronizing), and then hops to the acknowledgment frequency on which the receiver sends an ACK. If the data is large and requires multiple time slots, the sender indicates this in the header of the data packet. The receiver then sends an HR packet in the HR period of the next time slot, to extend the reservation of the current frequency for the sender and receiver. This tells the other nodes to skip this frequency in the hopping sequence. HRMA achieves significantly higher throughput than Slotted ALOHA in FHSS channels. It uses simple, half-duplex, slow-frequency-hopping radios that are commercially available. It requires, however, synchronization among nodes, which is not suitable for multihop networks.

2.3.4.4   Multichannel Medium Access Control (MMAC)

MMAC utilizes multiple channels by switching among them dynamically. Although the IEEE 802.11 protocol has inherent support for multiple channels in DCF mode, it only utilizes one channel at present. The primary reason is that hosts with a single half-duplex transceiver can only transmit or listen to one channel at a time. MMAC is an adaptation to the DCF to use multiple channels. Similar to the Dynamic Power-Saving Mechanism (DPSM) scheme, time is divided into multiple fixed-time beacon intervals. The beginning of every interval has a small Ad Hoc Traffic Indication Message (ATIM) window. During this window, ATIM packets are exchanged among nodes so that they can coordinate the assignment of appropriate channels for use in the subsequent time slots of that interval. Unlike other multichannel protocols, MMAC needs only one transceiver. At the beginning of every beacon interval, every node synchronizes itself to all other nodes by tuning in to a common synchronization channel on which ATIM packets are exchanged. No data packet transmission is allowed during this period. Further, every node maintains a Preferred Channel List (PCL) that stores the usage of channels within its transmission range, and also allows for marking priorities for those channels. If a node has a data packet to send, it sends out an ATIM packet to the recipient that includes the sender's PCL. The receiver in turn compares the sender's PCL with that of its own and selects an appropriate channel for use. It then responds with an ATIM-ACK packet and includes the chosen channel in it. If the chosen channel is acceptable to the sender, it responds with an ATIM-RES (reservation) packet. Any node overhearing an ATIM-ACK or ATIM-RES packet updates its own PCL. Subsequently, the sender and receiver exchange RTS–CTS messages on the selected channel prior to data exchange. Otherwise, if the chosen channel is not suitable for the sender, it has to wait till the next beacon interval to try another channel. The performance of MMAC is better than that of IEEE 802.11 and Dynamic Channel Assignment (DCA) in terms of throughput. Also, it can be easily integrated with IEEE 802.11 Power-Saving Mechanism (PSM) mode while using a simple hardware. However, it has a longer packet delay than DCA. Moreover, it is not suitable for multihop ad hoc networks as it assumes that the nodes are synchronized.

2.3.4.5 Dynamic Channel Assignment with Power Control (DCA-PC)

DCA-PC is an extension of their DCA protocol that did not consider the issue of power control. It combines concepts of power control and multiple-channel medium access in the context of MANETs. The hosts are assigned channels dynamically, as and when they need them. Every node is equipped with two half-duplex transceivers, and the bandwidth is divided into a control channel and multiple data channels. One transceiver operates on the control channel to exchange control packets (using maximum power) for reserving the data channel, and the other switches between the data channels for exchanging data and acknowledgments (with power control). When a host needs a channel to talk to another, it engages in an RTS–CTS–RES exchange, where RES is a special reservation packet, indicating the appropriate data channel to be used. Every node keeps a table of power levels to be used when communicating with any other node. These power levels are calculated based on the RTS–CTS exchanges on the control channel. Because every node is always listening to the control channel, it can even dynamically update the power values based on the other control exchanges happening around it. Every node maintains a list with channel usage information. In essence, this list tells the node which channel its neighbor is using and the times of such usage. DCA-PC has been shown to achieve higher throughput than DCA. However, it is observed that when the number of channels is increased beyond a point, the effect of power control is less significant due to overloading of the control channel. In summary, DCA-PC is a novel attempt at solving dynamic channel assignment and power control issues in an integrated fashion.

2.3.5 Power-Aware or Energy-Efficient MAC Protocols

Because mobile devices are battery powered, it is crucial to conserve energy and utilize power as efficiently as possible. In fact, the issue of power conservation should be considered across all the layers of the protocol stack. The following principles may serve as general guidelines for power conservation in MAC protocols.

1. Collisions are a major cause of expensive retransmissions and should be avoided as far as possible.
2. The transceivers should be kept in standby mode (or switched off) whenever possible as they consume the most energy when in active mode.
3. Instead of using the maximum power, the transmitter should switch to a lower power mode that is sufficient for the destination node to receive the transmission.

The details of some selected schemes are discussed below.

2.3.5.1 Power-Aware Medium Access Control with Signaling (PAMAS)

The basic idea of PAMAS is that all the RTS–CTS exchanges are performed over the signaling channel and the data transmissions are kept separate over a data channel. While receiving a data packet, the destination node starts sending out a busy tone over the signaling channel. Nodes listen in on the signaling channel to deduce when it is optimal for them to power down their transceivers. Every node makes its own decision about whether to power off or not, such that there is no drop in the throughput. A node powers itself off if it has nothing to transmit and it realizes that its neighbor is transmitting. A node also powers off if at least one neighbor is transmitting and another is receiving at the same time. There are several rules to determine the length of a power-down state. This scheme can be used with other protocols like Floor Acquisition Multiple Access (FAMA). FAMA has also noted that the use of ACK and the transmission of multiple packets together will also enhance the performance of PAMAS. However, the radio transceiver turnaround time, which might not be negligible, was not considered in the PAMAS scheme.

2.3.5.2 Dynamic Power-Saving Mechanism (DPSM)

DPSM is based on the idea of using sleep and wake states for nodes to conserve power. It is a variation of the IEEE 802.11 scheme, in that it uses dynamically sized ATIM windows to achieve longer dozing times for nodes. The IEEE 802.11 DCF mode has a power-saving mechanism, in which time is divided into beacon intervals that are used to synchronize the nodes. At the beginning of each beacon interval, every node must stay awake for a fixed time called the "ATIM window." This window is used to announce the status of packets ready for transmission to any receiver nodes. Such announcements are made through ATIM frames, and they are acknowledged with ATIM-ACK packets during the same beacon interval. Figure 2.5 illustrates the mechanism. Earlier work shows that if the size of the

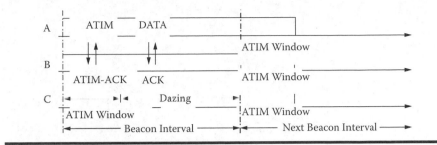

Figure 2.5   Power-saving mechanism for DCF.

ATIM window is kept fixed, performance suffers in terms of throughput and energy consumption.

In DPSM, each node dynamically and independently chooses the length of the ATIM window. As a result, every node can potentially end up having a differently sized window. It allows the sender and receiver nodes to go into sleep state immediately after they have participated in the transmission of packets announced in the prior ATIM frame. Unlike the DCF mechanism, they do not even have to stay awake for the entire beacon interval. The length of the ATIM window is increased if some packets queued in the outgoing buffer are still unsent after the current window expires. Also, each data packet carries the current length of the ATIM window, and any nodes that overhear such information may decide to modify their own window lengths based on the received information. DPSM is found to be more effective than IEEE 802.11 DCF in terms of power saving and throughput. However, IEEE 802.11 and DPSM are not suitable for multihop ad hoc networks as they assume that the clocks of the nodes are synchronized and the network is connected.

Node A announces a buffered packet for B using an ATIM frame. Node B replies by sending an ATIM-ACK, and both A and B stay awake during the entire beacon interval. The actual data transmission from A to B is completed during the beacon interval; because C does not have any packet to send or receive, it dozes after the ATIM window.

2.3.5.3 Power Control Medium Access Control (PCM)

Previous approaches of power control used alternating sleep and wake states for nodes. In PCM, the RTS and CTS packets are sent using the maximum available power, whereas the data and ACK packets are sent with the minimum power required to communicate between the sender and receiver. An example scenario is depicted in Figure 2.6. Node D sends the RTS to node E at a transmit power level P_{max}, and also includes this value in the packet. E measures the actual signal strength, say P_r, of the received RTS packet. Based on P_{max}, P_r, and the noise level at its location, E then computes the minimum necessary power level (say, P_{suff})

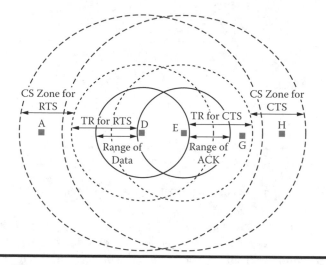

Figure 2.6 Illustration of power control scheme: (CS) carrier sense and (TR) transmission range.

that would actually be sufficient for use by D. Now, when E responds with the CTS packet using the maximum power it has available, it includes the value of P_{suff} that D subsequently uses for data transmission. G is able to hear this CTS packet and defers its own transmissions. E also includes the power level that it used for the transmission in the CTS packet. D then follows a similar process and calculates the minimum required power level that would get a packet from E to itself. It includes this value in the data packet so that E can use it for sending the ACK. PCM also stipulates that the source node periodically transmits the data packet at the maximum power level for just enough time so that nodes in the carrier-sensing range, such as A, may sense it. PCM thus achieves energy savings without causing throughput degradation. The operation of the PCM scheme requires a rather accurate estimation of received packet signal strength. Therefore, the dynamics of wireless signal propagation due to fading and shadowing effect may degrade its performance. Another drawback of this scheme is the difficulty in implementing frequent changes in the transmit power levels.

2.3.5.4 Power-Controlled Multiple Access (PCMA)

PCMA relies on controlling transmission power of the sender so that the intended receiver is just able to decipher the packet. This helps in avoiding interference with other neighboring nodes that are not involved in the packet exchange. PCMA uses two channels, one for sending out busy tones and the other for data and other control packets. The power control mechanism in PCMA has been used for increasing channel efficiency through spatial frequency reuse rather than only increasing battery life. Therefore, an important issue is for the transmitter and receiver pair to determine the minimum power level necessary for the receiver to decode the packet, while distinguishing it from noise or interference. Also, the receiver has to advertise its noise tolerances so that no other potential transmitter will disrupt its ongoing reception. In the conventional methods of collision avoidance, a node is either allowed to transmit or not, depending on the result of carrier sensing. In PCMA, this method is generalized to a bounded power model. Before data transmission, the sender sends a Request_Power_To_Send (RPTS) packet on the data channel to the receiver. The receiver responds with an Accept_Power_To_Send (APTS) packet, also on the data channel. This RPTS–APTS exchange is used to determine the minimum transmission power level that will cause a successful packet reception at the receiver. After this exchange, the actual data is transmitted and acknowledged with an ACK packet. In a separate channel, every receiver sets up a special busy tone as a periodic pulse. The signal strength of this busy tone advertises to the other nodes the additional noise power the receiver node can tolerate. When a sender monitors the busy tone channel, it is essentially doing something similar to carrier sensing, as in the CSMA/CA model. When a receiver sends out a busy tone pulse, it is doing something similar to sending out a CTS packet. The RPTS–APTS exchange is analogous to the RTS–CTS exchange. The major difference, however, is that the RPTS–APTS exchange does not force other hidden transmitters to back off. Collisions are resolved by the use of some appropriate backoff strategy.

2.4 Summary

This chapter focused on ad hoc wireless networks with respect to MAC protocols. Many schemes and their salient features were discussed. In particular, it concentrated on issues of collision resolution, power conservation, multiple channels, and advantages of using directional antennas. The characteristics and operating principles of several MAC schemes were discussed. Although some of them are general purpose protocols (such as MACA and MACAW), others focus on specific features such as power control (e.g., PAMAS and PCM) or the use of specialized technology like directional antennas (e.g., directional MAC [D-MAC] and Multihop RTS MAC [M-MAC]). Most of these schemes, however, are not designed specially for networks with mobile nodes. On the other hand, the transaction time at the MAC layer is relatively short. The effect of mobility will become less significant as the available channel bandwidth continues to grow.

2.5 Problems

2.1 Explain the need of MAC protocols for ad hoc networks.
2.2 Discuss the various factors that need to be considered while measuring the performance of the MAC protocol for ad hoc networks.

2.3 With a neat diagram, explain hidden- and exposed-node problems.

2.4 Give the classification of MAC protocols with a suitable example.

2.5 Describe contention-based MAC protocols with an example.

2.6 Explain contention-based MAC protocols with reservations.

2.7 Describe the MACA protocol with a suitable example.

2.8 Explain briefly the two modes of operation in the IEEE 802.11 MAC scheme.

2.9 Discuss the operation of Multiple Access Collision Avoidance—by Invitation (MACA-BI) with an illustration.

2.10 How would Group Allocation Multiple Access with Packet Sensing (GAMA-PS) work well with ad hoc networks? Explain.

2.11 Describe MAC protocols using directional antennas in detail.

2.12 Explain the advantage of multiple-channel MAC protocols.

2.13 Discuss the advantage of power-aware or energy-efficient MAC protocols.

2.14 Explain Power-Aware Medium Access Control with Signaling (PAMAS) with an example.

2.15 Explain how the Dynamic Power-Saving Mechanism (DPSM) is an efficient MAC protocol in ad hoc networks.

2.16 Discuss Power Control Medium Access Control (PCM) and Power-Controlled Multiple Access (PCMA).

Bibliography

N. Abramson, The ALOHA system: Another alternative for computer communications, Proc. Fall Joint Computer Conf., 37:281–285, 1970.

I. F. Akyildiz, J. McNair, L. C. Martorell, R. Puigjaner, and Y. Yesha, Medium access control protocols for multimedia traffic in wireless networks, *IEEE Network*, 13:39–47, July–Aug. 1999.

V. Bhargavan, A. Demers, S. Shenker, and L. Zhang, MACAW: A media access protocol for wireless LANs, *Proc. ACM SIGCOMM*, 1994, pp. 212–225.

S. Chakrabarti and A. Mishra, QoS issues in ad hoc wireless networks, *IEEE Commun.*, 39(2):142–148, Feb. 2001.

B. Chen, K. Jamieson, H. Balakrishnan, and R. Morris, Span: An energy-efficient coordination algorithm for topology maintenance in ad hoc wireless networks, ACM MOBICOM, July 2001.

K. C. Chen, Medium access protocols of wireless LANs for mobile computing, *IEEE Network*, 8(5):50–63, 1994.

A. Chockalingam and M. Zorzi, Energy efficiency of media access protocols for mobile data networks, *IEEE Trans. Commun.*, 46(11):1418–1421, 1998.

B. P. Crow, I. Widjaja, J. G. Kim, and P. T. Sakai, IEEE 802.11 wireless local area networks, IEEE Commun., Sept. 1997.

C. L. Fullmer and J. J. Garcia-Luna-Aceves, Floor acquisition multiple access (FAMA) for packet-radio networks, *Proc. ACM SIGCOMM*, Cambridge, MA, Aug. 28–Sep. 1, 1995.

R. G. Gallager, A perspective on multi access channels, *IEEE Trans. Inf. Th.*, 31(2):124–142, 1985.

A. J. Goldsmith and S. B. Wicker, Design challenges for energy-constrained ad hoc wireless networks, *IEEE Wireless Commun.*, 9(4):8–27, 2002.

Z. J. Haas and J. Deng, Dual busy tone multiple access (DBTMA): A multiple access control scheme for ad hoc networks, *IEEE Trans. Commun.*, 50(6):975–984, 2002.

Z. J. Haas and S. Tabrizi, On some challenges and design choices in ad-hoc communications, Proc. IEEE MILCOM98, Vol. 1, 1998.

IEEE 802.11 Working Group, Wireless LAN medium access control (MAC) and physical layer (PHY) specification, IEEE Press, Piscataway, NJ, 1997.

IETF MANET Working Group, http://www.ietf.org/html.charters/manet-charter.html.

E-S. Jung and N. H. Vaidya, A power control MAC protocol for ad hoc networks, ACM Intl. Conf. Mobile Computing and Networking (MOBICOM), Sept. 2002.

E-S. Jung and N. H. Vaidya, An energy efficient MAC protocol for wireless LANs, IEEE INFOCOM, June 2002.

P. Karn, MACA: A new channel access method for packet radio, ARRL/CRRL Amateur Radio 9th Computer Networking Conf., Sept. 22, 1990.

L. Kleinrock and F. A. Tobagi, Packet switching in radio channels: Part I—Carrier sense multiple access modes and their throughput-delay characteristics, *IEEE Trans. Commun.*, 23:1400–1416, Dec. 1975.

L. Kleinrock and F. A. Tobagi, Packet switching in radio channels: Part II—The hidden terminal problem in carrier sense multiple access and busy tone solution, *IEEE Trans. Commun.*, 23:1417–1433, Dec. 1975.

J. Monks, V. Bharghavan, and W. Hwu, A power controlled multiple access protocol for wireless packet networks, IEEE INFOCOM, April 2001.

A. Muir and J. J. Garcia-Luna-Aceves, An efficient packet sensing MAC protocol for wireless networks, *Mobile Networks and Applications*, 3(3):221–34, 1998.

B. O'Hara and A. Petrick, IEEE 802.11 Handbook: A designer's companion, IEEE Press, Piscataway, NJ, 1999.

C. E. Perkins, *Ad hoc networking*, Addison-Wesley, Reading, MA, 2001.

N. Poojary, S. V. Krishnamurthy, and S. Dao, Medium access control in a network of ad hoc mobile nodes with heterogeneous power capabilities, *Proc. IEEE ICC*, 3:872–877, 2001.

R. Ramanathan and R. R. Hain, Topology control of multihop wireless networks using transmit power adjustment, *IEEE INFOCOM*, Vol. 2, March 2000, pp. 404–413.

V. Rodoplu and T. H. Meng, Minimum energy mobile wireless networks, IEEE J. Sel. Areas Commun., 17(8):1338–1344, 1999.

E. M. Royer and C. K. Toh, A review of current routing protocols for ad hoc mobile wireless networks, *IEEE Personal Commun.*, 6(2):46–55, April 1999.

G. Sidhu, R. Andrews, and A. Oppenheimer, *Inside AppleTalk*, Addison-Wesley, Reading, MA, 1989.

S. Singh and C. S. Raghavendra, PAMAS: Power Aware Multi-Access protocol with signaling for ad hoc networks, *ACM Computer Commun. Rev.*, 28(3):5–26, 1998.

K. M. Sivalingam, M. B. Srivastava, and P. Agrawal, Low power link and access protocols for wireless multimedia networks, IEEE VTC, May 1997.

W. M. Smith and P. S. Ghang, A low power medium access control protocol for portable multi-media systems, 3rd Intl. Workshop Mobile Multimedia Commun., Sept. 25–27, 1996.

F. Talucci and M. Gerla, MACA-BI (MACA by Invitation): A wireless MAC protocol for high speed ad hoc networking, Proc. IEEE ICUPC, 1997.

Z. Tang and J. J. Garcia-Luna-Aceves, Hop-reservation multiple access (HRMA) for ad-hoc networks, IEEE INFOCOM, 1999.

C-K. Toh, *Ad hoc mobile wireless networks: Protocols and systems*, Prentice Hall PTR, Upper Saddle River, NJ, 2002.

Y-C. Tseng, C-S. Hsu, and T-Y. Hsieh, Power-saving protocols for IEEE 802.11-based multi-hop ad hoc networks, Proc. IEEE INFOCOM, 2002.

R. Wattenhofer, L. Li, P. Bahl, and Y. M. Wang, Distributed topology control for power efficient operation in multihop wireless ad hoc networks, *IEEE INFOCOM*, Vol. 3, April 2001, pp. 1388–1397.

H. Woesner, J. P. Ebert, M. Schlager, and A. Wolisz, Power saving mechanisms in emerging standards for wireless LANs: The MAC level perspective, *IEEE Personal Commun.*, 5(3):40–48, 1998.

C. Wu and V. O. K. Li, Receiver-initiated busy-tone multiple access in packet radio networks, *Proc. ACM SIGCOMM Conf.*, 1987, pp. 336–342.

W. Ye, J. Jeidemann, and D. Estrin, An energy-efficient MAC protocol for wireless sensor networks, Proc. IEEE INFOCOM, 2002.

Chapter 3

Routing Protocols for Ad Hoc Wireless Networks

3.1 Introduction

With the advances of wireless communication technology, low-cost and powerful wireless transceivers are widely used in mobile applications. Mobile networks have attracted significant interest in recent years because of their improved flexibility and reduced costs. Compared to wired networks, mobile networks have unique characteristics. In mobile networks, node mobility may cause frequent network topology changes, which are rare in wired networks. In contrast to the stable link capacity of wired networks, wireless link capacity continually varies because of the impacts from transmission power, receiver sensitivity, noise, fading, and interference. Additionally, wireless mobile networks have a high error rate, power restrictions, and bandwidth limitations.

Mobile networks can be classified into infrastructure networks and mobile ad hoc networks (MANETs) according to their dependence on fixed infrastructures. In an infrastructure mobile network, mobile nodes have wired access points (or base stations) within their transmission range. The access points compose the backbone for an infrastructure network. In contrast, mobile ad hoc networks are autonomously self-organized networks without infrastructure support. In a mobile ad hoc network, nodes move arbitrarily; therefore the network may experience rapid and unpredictable topology changes. Additionally, because nodes in a mobile ad hoc network normally have limited transmission ranges, some nodes cannot communicate directly with each other. Hence, routing paths in mobile ad hoc networks potentially contain multiple hops, and every node in mobile ad hoc networks has the responsibility to act as a router.

Mobile ad hoc networks originated from the U.S. Government's Defense Advanced Research Projects Agency (DARPA) Packet Radio Network (PRNet) and SURAN project. Being independent on preestablished infrastructure, mobile ad hoc networks have advantages such as rapidity and ease of deployment, improved flexibility, and reduced costs. Mobile ad hoc networks are appropriate for mobile applications in either hostile environments where no infrastructure is available, or temporarily established mobile applications, which are cost crucial. In recent years, application domains of mobile ad hoc networks have gained more and more importance in nonmilitary public organizations and in commercial and industrial areas. The typical application scenarios include rescue missions, law enforcement operations, cooperating industrial robots, traffic management, and educational operations in campus.

Active research work for mobile ad hoc networks is carrying on mainly in the fields of Medium Access Control (MAC), routing, resource management, power control, and security. Because of the importance of routing protocols in dynamic multihop networks, a lot of mobile ad hoc network routing protocols have been proposed in the last few years. There are some challenges that make the design of mobile ad hoc network routing protocols a tough task. Firstly, in mobile ad hoc networks, node mobility causes frequent topology changes and network partitions. Secondly, because of the variable and unpredictable capacity of wireless links, packet losses may happen frequently. Moreover, the broadcast nature of the wireless medium introduces the hidden terminal and exposed terminal problems. Additionally, mobile nodes have restricted power, computing, and bandwidth resources and require effective routing schemes. As promising network types in future mobile applications, mobile ad hoc networks are attracting more and more researchers. This chapter gives the state-of-the-art review for typical routing protocols for mobile ad hoc networks, including classical MANET unicast and multicast routing algorithms and popular classification methods.

3.2   Design Issues of Routing Protocols for Ad Hoc Networks

The major design issues in mobile ad hoc networks are discussed below.

3.2.1   Routing Architecture

The routing architecture of a self-organized network can be either hierarchical or flat. In most self-organized networks, the hosts will be acting as independent routers, which implies that routing architecture should conceptually be flat, that is, each address serves only as an identifier and does not convey any information about one host that is topologically located with respect to any other node. In a flat self-organized network, the mobility management is not necessary because all of the nodes are visible to each other via routing protocols. In flat routing algorithms such as Destination-Sequenced Distance Vector (DSDV) and Wireless Routing Protocol (WRP), the routing tables have entries to all hosts in the self-organized network. However, a flat routing algorithm does not have good scalability. The routing overhead increases rapidly when the network becomes larger. Hence, to control channel reuse spatially (in terms of frequency, time, or spreading code) and reduce routing information overhead, some form of hierarchical scheme should be employed. Clustering is the most common technique employed in hierarchical routing architectures. The idea behind hierarchical routing is to divide the hosts of a self-organized network into a number of overlapping or disjoint clusters. One node is elected as clusterhead for each cluster. This clusterhead maintains the membership information for the cluster.

Nodes that are not clusterheads will, henceforth, be referred to as "ordinary nodes." When an ordinary node wants to send a packet, the node can send the packet to the clusterhead that routes the packet toward the destination. Clusterhead Gateway Switch Routing (CGSR) and Cluster-Based Routing Protocol (CBRP) belong to this type of routing scheme. Hierarchical routing involves cluster, address, and mobility management.

3.2.2 Unidirectional Links Support

Almost every existing routing protocol tends to assume that all links are bidirectional. However, there are a number of factors that will make wireless links unidirectional. They are as follows:

- *Different radio capabilities*: Radios within a network can have different transmit powers or receive sensitivities. This is quite likely in a tactical environment where man-pack and vehicular radios exist. Vehicular radios, being less constrained by size and weight, typically have 12 decibels (dB) greater transmit power than their man-pack counterparts. Unidirectional links are exceedingly likely in tactical networks.
- *Interference*: This is due to either hostile jammers or friendly interference, which will reduce a nearby receiver's sensitivity. For example, host A can receive packets from host B as there is very little interference in A's vicinity. However, B may be in the vicinity of an interference node, and therefore cannot receive packets from A. So, the link between A and B is directed from B to A.

- *Message broadcast requirement*: There is an increasing emphasis on the wide area broadcast of messages. Satellite-based transmitters are being used for the downward links, whereas the upward links use alternative paths.
- *Mute mode*: An extreme instance, applicable only in tactical mobile networks, is when hosts cannot transmit due to an impending threat. In such a case, it still needs to receive information; however, it cannot participate in bidirectional communications.
- *The state of link direction is time varying*: The directional state of the wireless link may be either a persistent or a transient phenomenon. The frequency of such transitions and the duration of stay in each state will be a function of offered traffic, terrain, mobility patter, and energy availability.

3.2.3 Usage of SuperHosts

All existing routing protocols assume that all mobile hosts have the same properties based on the spirit of a self-organized network as a collection of "equal" peers opportunistically using each other's services to communicate. Although this is true in some circumstances, there are also situations where the network will include hosts with preponderant bandwidth, guaranteed power supply, and high-speed wireless links. Such hosts are referred to as "SuperHosts." For example, a company in a military environment consists of a number of walking soldiers equipped with low-capacity man-pack radios and a few tanks with high-capacity vehicular radios. Usually, the self-organized networks in this situation have two-level network architecture: backbone area and subarea. Backbone area is composed of SuperHosts. In addition, SuperHosts are often assumed to have lower mobility than normal hosts so as to maintain the stability of the backbone. Normal hosts need not make routing decisions. For example, a satellite host (a SuperHost) can easily collect the routing information from the normal hosts' geographical locations, build the routing table, and propagate these routes. The example is just analogous that a person on stage is likely to have a much better view of the wireless network throughout an auditorium.

3.2.4 Quality of Service (QoS) Routing

Up to now, most of the routing protocols that have been proposed for ad hoc wireless networks optimized the solution for only one metric: hop distance. So the shortest path is generally preferable. For datagram traffic, shortest path routing may be sufficient. However, these wireless links in self-organized networks, typically scarce and dynamic, make it difficult to perform efficient resource utilization or to execute critical real-time applications in such environments. Based on this consideration, it is necessary to provide QoS routing support to effectively control the total traffic that can flow into the network. QoS routing is a routing mechanism under which paths for flows are determined according to resource availability in the network as well as the QoS requirement of flows. QoS routing means that it selects routes with sufficient resources for the requested QoS parameters. The goal of QoS routing has two points. The first one is to meet the QoS requirements for each admitted connection, and the second one is to achieve global efficiency in resource utilization. Thus, QoS routing will consider multiple constraints, and provide better load balance by allocating traffic on different paths, subject to the QoS requirement of different traffic. On the contrary, current routing protocols seem to favor routing traffic based on the shortest path, thereby causing a bottleneck. In a self-organized network, there are many metrics to be considered: the (1) most reliable path, (2) most stable path, (3) maximum total power remained path, (4) maximum available bandwidth path, and so forth. It is desirable to select the routes with a minimum cost based on the above metrics and not unlikely only to provide the shortest path based on the hop distance.

3.2.5 Multicast Support

As we know, multicast routing is a network-layer function that constructs paths along which data packets from a source are distributed to reach many, but not all, destinations in a communication network. Then, multicast routing sends a single copy of a data packet simultaneously to multiple receivers over a communication link that is shared by the paths to the receivers. The sharing of links in the collection of the paths to receivers implicitly defines a tree used to distribute multicast packets. In contrast to unicast routing, multicast routing is a very useful and efficient way to support group communication. This is especially the case in self-organized networks where bandwidth is limited and energy is constrained. In addition, a self-organized network often consists of several cooperative work groups. The deployment of multicast routing in a self-organized network will provide collaborative visualization and multimedia conferencing as well as information dissemination in critical situations such as disaster or military scenarios. Multicast routing in self-organized networks became an active research topic only very recently, and much research has focused on designing the Unicast Routing Protocols. However, a self-organized network is better suited for multicast than unicast because of its broadcast characteristics.

Employing multicast routing in a self-organized network poses new challenges. Traditional multicast protocols are not suitable for this environment because of the following reasons:

1. The source-originating route requests a move, making source-oriented protocols inefficient.
2. Multicast members move, thus precluding the use of fixed multicast topology.
3. Transient loops may form during spanning tree reconfiguration.

4. Maintaining too much multicast-related state information puts much pressure on both storage capacity and power, which are severely limited in handheld devices in self-organized networks.

3.3 Classification of Routing Protocols

The limited resources in MANETs have made designing an efficient and reliable routing strategy a very challenging problem. An intelligent routing strategy is required to efficiently use the limited resources while at the same time be adaptable to the changing network conditions such as network size, traffic density, and network partitioning. In parallel with this, the routing protocol may need to provide different levels of QoS to different types of applications and users.

Prior to the increased interests in wireless networking, in wired networks two main algorithms were used. These algorithms are commonly referred to as the link-state and distance-vector algorithms. In link-state routing, each node maintains an up-to-date view of the network by periodically broadcasting the link-state costs of its neighboring nodes to all other nodes using a flooding strategy. When each node receives an update packet, it updates its view of the network and its link-state information by applying a shortest-path algorithm to choose the next-hop node for each destination. The traditional link-state and distance-vector algorithms do not scale in large MANETs. This is because periodic or frequent route updates in large networks may consume a significant part of the available bandwidth, increase channel contention, and require each node to frequently recharge its power supply. To overcome the problems associated with the link-state and distance-vector algorithms, a number of routing protocols have been proposed for MANETs. These protocols can be classified into three different groups: global or proactive, on demand or reactive, and hybrid. In proactive routing protocols, the routes to all the destinations (or parts of the network) are determined at the start-up and maintained by using a periodic route update process. In reactive protocols, routes are determined when they are required by the source using a route discovery process. Hybrid routing protocols combine the basic properties of two classes of protocols into one. That is, they are both reactive and proactive in nature. Each group has a number of different routing strategies, which employ a flat or a hierarchical routing structure.

To compare and analyze mobile ad hoc network routing protocols, appropriate classification methods are important. Classification methods help researchers and designers to understand distinct characteristics of a routing protocol and find its relationship with others. These characteristics mainly are related to the information which is exploited for routing, when this information is acquired, and the roles which nodes may take in the routing process.

3.3.1 Proactive, Reactive, and Hybrid Routing

One of the most popular methods to distinguish mobile ad hoc network routing protocols is based on how routing information is acquired and maintained by mobile nodes. Using this method, mobile ad hoc network routing protocols can be divided, as discussed above, into proactive routing, reactive routing, and hybrid routing.

A proactive routing protocol is also called a "table-driven" routing protocol. Using a proactive routing protocol, nodes in a mobile ad hoc network continuously evaluate routes to all reachable nodes and attempt to maintain consistent, up-to-date routing information. Therefore, a source node can get a routing path immediately if it needs one.

In proactive routing protocols, all nodes need to maintain a consistent view of the network topology. When a network topology change occurs, respective updates must be propagated throughout the network to notify the change. Most proactive routing protocols proposed for mobile ad hoc networks have inherited properties from algorithms used in wired networks. To adapt to the dynamic features of mobile ad hoc networks, necessary modifications have been made on traditional wired network routing protocols. Using proactive routing algorithms, mobile nodes proactively update the network state and maintain a route regardless of whether data traffic exists or not, and the overhead to maintain up-to-date network topology information is high. The next section will introduce several typical proactive mobile ad hoc network routing protocols, such as the WRP, DSDV, and the Fisheye State Routing (FSR) Protocols.

Reactive routing protocols for mobile ad hoc networks are also called "on-demand" routing protocols. In a reactive routing protocol, routing paths are searched only when needed. A route discovery operation invokes a route-determination procedure. The discovery procedure terminates when either a route has been found or no route is available after examination for all route permutations.

In a mobile ad hoc network, active routes may be disconnected due to node mobility. Therefore, route maintenance is an important operation of reactive routing protocols. Compared to the proactive routing protocols for mobile ad hoc networks, less control overhead is a distinct advantage of the reactive routing protocols. Thus, reactive routing protocols have better scalability than proactive routing protocols in mobile ad hoc networks. However, when using reactive routing protocols, source nodes may suffer from long delays for route searching before they can forward data packets. The Dynamic Source Routing (DSR) Protocol and Ad Hoc On-Demand Distance Vector (AODV) Routing Protocol are examples of reactive routing protocols for mobile ad hoc networks.

Hybrid routing protocols are proposed to combine the merits of both proactive and reactive routing protocols and overcome their shortcomings. Normally, hybrid routing protocols for mobile ad hoc networks exploit hierarchical network architectures. The proper proactive routing approach and reactive routing approach are exploited in different hierarchical levels, respectively. In this chapter, as examples of hybrid routing protocols for mobile ad hoc networks, the Zone Routing Protocol (ZRP), Zone-Based Hierarchical Link State (ZHLS) Routing Protocol, and Hybrid Ad Hoc Routing Protocol (HARP) will be introduced and discussed.

3.3.2 Structuring and Delegating the Routing Task

Another classification method is based on the roles which nodes may have in a routing scheme. In a uniform routing protocol, all mobile nodes have the same role, importance, and functionality. Examples of uniform routing protocols include the WRP, DSR, AODV, and DSDV Routing Protocols. Uniform routing protocols normally assume a flat network structure. In a nonuniform routing protocol for mobile ad hoc networks, some nodes carry out distinct management or routing functions. Normally, distributed algorithms are exploited to select those special nodes. In some cases, nonuniform routing approaches are related to hierarchical network structures to facilitate node organization and management. Nonuniform routing protocols further can be divided according to the organization of mobile nodes and how management and routing functions are performed. Following these criteria, nonuniform routing protocols for mobile ad hoc networks are divided into zone-based hierarchical routing, cluster-based hierarchical routing, and core-node-based routing.

In zone-based routing protocols, different zone-constructing algorithms are exploited for node organization; for example, some zone-constructing algorithms use geographical information. Exploiting zone division effectively reduces the overhead for routing information maintenance. Mobile nodes in the same zone know how to reach each other with smaller cost compared to maintaining routing information for all nodes in the whole network. In some zone-based routing protocols, specific nodes act as gateway nodes and carry out interzone communication. The ZRP and the ZHLS Routing Protocol are zone-based hierarchical routing protocols for mobile ad hoc networks.

A cluster-based routing protocol uses a specific clustering algorithm for cluster-head election. Mobile nodes are grouped into clusters, and clusterheads take the responsibility for membership management and routing functions. Clusterhead Gateway Switch Routing (CGSR) will be introduced in a later section of this chapter as an example of a cluster-based mobile ad hoc network routing protocol. Some cluster-based mobile ad hoc network routing protocols potentially support a multilevel cluster structure, such as the Hierarchical State Routing (HSR).

In core-node-based routing protocols for mobile ad hoc networks, critical nodes are dynamically selected to compose a "backbone" for the network. The "backbone" nodes carry out special functions, such as routing path construction and control and data packets propagation. Core-Extraction Distributed Ad Hoc Routing (CEDAR) is a typical core-node-based mobile ad hoc network routing protocol.

3.3.3 Exploiting Network Metrics for Routing

Metrics used for routing path construction can be used as criteria for mobile ad hoc network routing protocol classification. Most routing protocols for mobile ad hoc networks use "hop number" as a metric. If there are multiple routing paths available, the path with the minimum hop number will be selected. If all wireless links in the network have the same failure probability, short routing paths are more stable than the long ones and can obviously decrease traffic overhead and reduce packet collisions. However, the assumption of the same failure properties may not be true in mobile ad hoc networks. Therefore, the stability of a link has to be considered in the route construction phase. For example, routing approaches such as Associatively Based Routing (ABR) and Signal Stability-Based Routing (SSR) are proposed that use link stability or signal strength as a metric for routing.

With the popularity of mobile computing, some mobile applications may have different QoS requirements. To meet specific QoS requirements, appropriate QoS metrics should be used for packet routing and forwarding in mobile ad hoc networks. As in wired networks, QoS routing protocols for mobile ad hoc networks can use metrics, such as bandwidth, delay, delay jitter, packet loss rate, and cost. As an example, bandwidth and link stability are used in CEDAR as metrics for routing path construction.

3.3.4 Evaluating Topology, Destination, and Location for Routing

In a topology-based routing protocol for mobile ad hoc networks, nodes collect network topology information for making routing decisions. Other than topology-based routing protocols, there are some destination-based routing protocols proposed in mobile ad hoc networks. In a destination-based routing protocol, a node only needs to know the next hop along the routing path when forwarding a packet to the destination. For example, DSR is a topology-based routing protocol, and AODV and DSDV are destination-based routing protocols. The availability of a Global Positioning System (GPS) or similar locating systems allows mobile nodes to access geographical information easily. In location-based routing protocols, the position relationship between a packet-forwarding node and the destination, together with the node mobility, can be used in both route discovery and packet forwarding. Existing location-based routing approaches for mobile ad hoc networks can be divided into two schemes. In the first scheme, mobile nodes send packets merely depending on the location information and do not need any extra knowledge. The other scheme uses both location information and topology information. Location-Aided Routing (LAR) and Distance Routing Effect Algorithm for Mobility (DREAM) are typical location-based routing protocols proposed for mobile ad hoc networks.

3.4 Proactive Routing Protocols

A proactive routing protocol is, as mentioned, also called a "table-driven" routing protocol. Using a proactive routing protocol, nodes in a mobile ad hoc network continuously evaluate routes to all reachable nodes and attempt to maintain consistent, up-to-date routing information. Therefore, a source node can get a routing path immediately if it needs one.

In proactive routing protocols, all nodes need to maintain a consistent view of the network topology. When a network topology change occurs, respective updates must be propagated throughout the network to notify the change. Most proactive routing protocols proposed for mobile ad hoc networks have inherited properties from algorithms used in wired networks. To adapt to the dynamic features of mobile ad hoc networks, necessary modifications have been made on traditional wired network routing protocols. Using proactive routing algorithms, mobile nodes proactively update the network state and maintain a route regardless of whether data traffic exists or not; the overhead to maintain up-to-date network topology information is high. In this section, we will discuss several typically proactive mobile ad hoc network routing protocols, such as the following:

- WRP
- DSDV
- Optimized Link State Routing (OLSR) Protocol
- FSR
- HSR
- Topology Broadcast Reverse Forwarding (TBRF)

3.4.1 Wireless Routing Protocol (WRP)

WRP is a proactive unicast routing protocol for mobile ad hoc networks. WRP uses an improved Bellman-Ford Distance Vector routing algorithm. To adapt to the dynamic features of mobile ad hoc networks, some mechanisms are introduced to ensure the reliable exchange of update messages and reduced route loops.

Using WRP, each mobile node maintains a distance table, a routing table, a link-cost table, and a Message Retransmission List (MRL). An entry in the routing table contains the distance to a destination node, the predecessor and the successor along the paths to the destination, and a tag to identify its state (i.e., is it a simple path, a loop, or invalid?). Storing predecessor and successor in the routing table helps to detect routing loops and avoid the counting-to-infinity problem, which is the main shortcoming of the original distance vector routing algorithm. A mobile node creates an entry for each neighbor in its link-cost table. The entry contains the cost of the link connecting to the neighbor and the number of timeouts because an error-free message was received from that neighbor.

In WRP, mobile nodes exchange routing tables with their neighbors using update messages. The update messages can be sent either periodically or whenever link state changes happen. The MRL contains information about which neighbor has not acknowledged an update message. If needed, the update message will be retransmitted to the neighbor. Additionally, if there is no change in its routing table since the last update, a node is required to send a Hello message to ensure connectivity. On receiving an update message, the node modifies its distance table and looks for better routing paths according to the updated information.

In WRP, a node checks the consistency of its neighbors after detecting any link change. A consistency check helps to eliminate loops and speed up convergence. One shortcoming of WRP is that it needs large memory storage and computing resources to maintain several tables. Moreover, as a proactive routing protocol, it has a limited scalability and is not suitable for large mobile ad hoc networks.

3.4.1.1 Overview

To describe WRP, we model a network as an undirected graph represented as G (V, E), where V is the set of nodes and E is the set of links (or edges) connecting the nodes. Each node represents a router and is a computing unit involving a processor, local memory, and input and output queues with unlimited capacity. In a wireless network, a node has radio connectivity with multiple nodes, and a single physical radio link connects a node with many other nodes. However, for the purposes of routing-table updating, a node A can consider another node B to be adjacent (we call such a node a "neighbor") if there is radio connectivity between A and B and A receives update messages from B. Accordingly, we map a physical broadcast link connecting multiple nodes into multiple point-to-point functional links defined for these node paths that are considered to be neighbors of each other.

Then, a functional bidirectional link connecting the nodes is assigned a positive weight in each direction. All messages received (transmitted) by a node are put in an input (output) queue and are processed in First In–First Out (FIFO) order. The communication links in the network are such that all update messages transmitted over an operational link are received in the order in which they were transmitted within a finite time.

A link is assumed to exist between two nodes only if there is radio connectivity between the two nodes and they can exchange update messages reliably with a certain probability of success. When a link fails, the corresponding distance entries in a node's distance and routing tables are marked as infinity. A node failure is modeled as all links incident on that node failing at the same time.

WRP is designed to run on top of the MAC protocol of a wireless network. Update messages may be lost or corrupted due to changes in radio connectivity or jamming. Reliable transmission of update messages is implemented by means of retransmissions. After receiving an update message free of errors, a node is required to send a positive acknowledgment (ACK) packet indicating that it has good radio connectivity and has processed the update message.

Because of the broadcast nature of the radio channel, a node can send a single update message to inform all its neighbors about changes in its routing table; however, each such neighbor sends an ACK to the originator node. In addition to ACKs, the connectivity can also be ascertained with the receipt of any message from a neighbor (which need not be an update message). To ensure that connectivity with a neighbor still exists when there are no recent transmissions of routing table updates or ACKs, periodic update messages without any routing table changes (null update messages) are sent to the neighbors. The time interval between two such null update messages is the HelloInterval. If a node fails to receive any type of message from a neighbor for a specified amount of time (e.g., three or four times the HelloInterval known as the Router-DeadInterval), the node must assume that connectivity with that neighbor has been lost.

3.4.1.2 Information Maintained at Each Node

For the purpose of routing, each node maintains a distance table, a routing table, a link-cost table, and an MRL.

The distance table of node i is a matrix containing, for each destination j and each neighbor of i (say, k), the distance to j (D^i_{jk}) and the predecessor (p^i_{jk}) reported by k.

The routing table of a node i is a vector with an entry for each known destination j, which specifies the following:

- The destination's identifier.
- The distance to the destination (D^i_j).
- The predecessor of the chosen shortest path to j (p^i_j).
- The successor (s^i_j) of the chosen shortest path to j.
- A marker (tag^i_j) used to update the routing table; it specifies whether the entry corresponds to a simple path (tag^i_j = correct), a loop (tag^i_j = error), or a destination that has not been marked (tag^i_j = null).

The link-cost table of node i lists the cost of relaying information through each neighbor k, and the number of update periods that have elapsed since node i received any error-free messages from k. The cost of a failed link is considered to be infinity.

The cost of a link could simply be 1, reflecting the hop count, or the addition of the latency over the link plus some constant bias. The cost of the link from i to k (i, k) is denoted by l^i_k.

The MRL specifies one or more retransmission entries, where the mth entry consists of the following:

- The sequence number of an update message.
- A retransmission counter that is decremented every time node i sends a new update message.
- An ACK-required flag (denoted by a^i_{km}) that specifies whether node k has sent an ACK to the update message represented by the retransmission entry.
- The list of updates sent in the update message.

The above information permits node i to know which updates of an update message (each update message contains a list of updates) have to be retransmitted and which neighbors should be requested to acknowledge such retransmission.

Node i retransmits the list of updates in an update message when the retransmission counter of the corresponding entry in the MRL reaches zero. The retransmission counter of a new entry in the MRL is set equal to a small number (e.g., 3 or 4).

3.4.1.3 Information Exchanged among Nodes

In WRP, nodes exchange routing-table update messages (which we call "update messages" for brevity) that propagate only from a node to its neighbors. An update message contains the following information:

- The identifier of the sending node.
- A sequence number assigned by the sending node.
- An update list of zero or more updates or ACKs to update messages. An update entry specifies a destination, a distance to the destination, and a predecessor to the destination. An ACK entry specifies the source and sequence number of the update message being acknowledged.
- A response list of zero or more nodes that should send an ACK to the update message.

In the event that the message space is not large enough to contain all the updates and ACKs that a node wants to report, they are sent in multiple update messages. An example of this event can be the case in which a node identifies a new neighbor and sends its entire routing table.

The response list of the update message is used to avoid the situation in which a neighbor is asked to send multiple ACKs to the same update message, simply because some other neighbor of the node sending the update did not acknowledge.

The first transmission of an update message must ask all neighbors to send an ACK, of course, and this is accomplished by specifying the all-neighbors address, which consists of all 1's. When the update message reports no updates, the empty address is specified; this address consists of all 0s and instructs the receiving nodes not to send an ACK in return. This type of update message is used as a hello message from a node to allow its neighbors to know that they maintain connectivity, even if no user messages or routing-table updates are exchanged. As we explain subsequently, an ACK entry refers to an entire update message, not an update entry in an update message, to conserve bandwidth.

3.4.1.4   Routing-Table Updating

A node can decide to update its routing table after either receiving an update message from a neighbor or detecting a change in the status of a link to a neighbor. When a node i receives an update message from its neighbor k, it processes each update and ACK entry of the update message in order. In WRP, a node checks the consistency of predecessor information reported by all its neighbors each time it processes an event involving a neighbor k. In contrast, all previous path-finding algorithms check the consistency of the predecessor only for the neighbor associated with the input event. This unique feature of WRP accounts for its fast convergence after a single resource failure or recovery as it eliminates more temporary looping situations than previous path-finding algorithms.

3.4.2   Destination-Sequenced Distance Vector (DSDV)

DSDV is a proactive unicast mobile ad hoc network routing protocol. Like WRP, DSDV is also based on the traditional Bellman-Ford algorithm. However, its mechanisms to improve routing performance in mobile ad hoc networks are quite different. In routing tables of DSDV, an entry stores the next hop toward a destination, the cost metric for the routing path to the destination, and a destination sequence number that is created by the destination. Sequence numbers are used in DSDV to distinguish stale routes from fresh ones and avoid the formation of route loops.

The route updates of DSDV can be either time driven or event driven. Every node periodically transmits updates, including its routing information, to its immediate neighbors. While a significant change occurs from the last update, a node can transmit its changed routing table in an event-triggered style. Moreover, the DSDV has two ways when sending routing table updates. One is the "full-dump" update type in which the full routing table is included inside the update. An incremental update, in contrast, contains only those entries with metrics that have been changed since the last update was sent. Additionally, the incremental update fits in one packet.

3.4.2.1   Distance Vector

In distance-vector algorithms, every node i maintains, for each destination z, a set of distances $\{d^x_{ij}\}$ where j ranges over the neighbors of i. Node i treats neighbor k as a next hop for a packet destined for x if d^x_{ik} equals $\min_j \{ d^x_{ij} \}$. The succession of next hops chosen in this manner leads to x along the shortest path. To keep the distance estimates up-to-date, each node monitors the cost of its outgoing links and periodically broadcasts, to each one of its neighbors, its current estimate of the shortest distance to every other node in the network.

The above distance-vector algorithm is the classical Distributed Bellman-Ford (DBF) algorithm. Compared to link state method, it is computationally more efficient, is easier to implement, and requires much less storage space; however, it is well-known that this algorithm can cause the formation of both short-lived and long-lived loops. The primary cause for formation of routing loops is that nodes choose their next hops in a completely distributed fashion based on information which can possibly be stale and, therefore, incorrect.

Almost all proposed modifications to the DBF algorithm eliminate the looping problem by forcing all nodes in the network to participate in some form of inter-nodal coordination protocol. Such internodal coordination mechanisms might be effective when topological changes are rare. However, within an ad hoc mobile environment, enforcing any such internodal coordination mechanism will be difficult due to the rapidly changing topology of the underlying routing network.

Simplicity is one of the primary attributes which makes any routing protocol preferred over others for implementation within operational networks. Routing Information Protocol (RIP) is a classical example. Despite the counting-to-infinity problem, it has proven to be very successful within small-size internetworks. The usefulness of RIP within an ad hoc environment, however, is limited as it was not designed to handle rapid topological changes. Furthermore, the techniques of split-horizon and poisoned-reverse are not useful within the wireless environment due to the broadcast nature of the transmission medium.

Packets are transmitted between the stations of the network by using routing tables that are stored at each station of the network. Each routing table, at each of the stations, lists all available destinations and the number of hops to each. Each route table entry is tagged with a sequence number which is originated by the destination station. To maintain the consistency of routing tables in a dynamically varying topology, each station periodically transmits updates, and transmits updates immediately when significant new information is available. Because we do not assume that the mobile hosts are maintaining any sort of time synchronization, we also make no assumption about the phase relationship of the update periods between the mobile hosts. These packets indicate which stations are accessible from each station and the number of hops necessary to reach these accessible stations, as is often done in distant-vector routing algorithms. It is not the purpose of this chapter to propose any new metrics for route selection other than the freshness of the sequence numbers associated with the route; cost or other metrics might easily replace the number of hops in other implementations. The packets may be transmitted containing either layer-2 (MAC) addresses or layer-3 (network) addresses.

Routing information is advertised by broadcasting or multicasting the packets which are transmitted periodically and incrementally as topological changes are detected (e.g., when stations move within the network). Data is also kept about the length of time between arrival of the first and the arrival of the best route for each particular destination. Based on this data, a decision may be made to delay advertising routes which are about to change soon, thus damping fluctuations of the route tables. The advertisement of routes which may not have stabilized yet is delayed to reduce the number of rebroadcasts of possible route entries that normally arrive with the same sequence number. The DSDV protocol requires each mobile station to advertise, to each of its current neighbors, its own routing table (for instance, by broadcasting its entries). The entries in this list may change fairly dynamically over time, so the advertisement must be made often enough to ensure that every mobile computer can almost always locate every other mobile computer of the collection.

In addition, each mobile computer agrees to relay data packets to other computers upon request. This agreement places a premium on the ability to determine the shortest number of hops for a route to a destination; we would like to avoid unnecessarily disturbing mobile hosts if they are in sleep mode. In this way a mobile computer may exchange data with any other mobile computer in the group even if the target of the data is not within range for direct communication. If the notification of which other mobile computers are accessible from any particular computer in the collection is done at layer 2, then DSDV will work with whatever higher layer (e.g., network-layer) protocol might be in use.

All the computers interoperating to create data paths between themselves broadcast the necessary data periodically, say, once every few seconds. In a wireless medium, it is important to keep in mind that broadcasts are limited in range by the physical characteristics of the medium. This is different than the situation with wired media, which usually have a much more well-defined range of reception. The data broadcast by each mobile computer will contain its new sequence number and the following information for each new route:

- The destination's address.
- The number of hops required to reach the destination.
- The sequence number of the information received regarding that destination, as originally stamped by the destination.

The transmitted routing tables will also contain the hardware address, and (if appropriate) the network address, of the mobile computer transmitting them, within the headers of the packet. The routing table will also include a sequence number created by the transmitter.

Routes with more recent sequence numbers are always preferred as the basis for making forwarding decisions because they might have a better metric. A mobile host could conceivably always receive two routes to the same destination, with a newer sequence number, one after another (via different neighbors), but always get the route with the worse metric first. Unless care is taken, this will lead to a continuing burst of new route transmittals upon every new sequence number from that destination. Each new metric is propagated to every mobile host in the neighborhood, which propagates to their neighbors, and so on. One solution is to delay the advertisement of such routes, when a mobile host can determine that a route with a better metric is likely to show up soon. The route with the later sequence number must be available for use, but it does not have to be advertised immediately unless it is a route to a destination which was previously unreachable. Thus, there will be two routing tables kept at each mobile host: one for use with forwarding packets, and another to be advertised via incremental routing information packets. To determine the probability of imminent arrival of routing information showing a better metric, the mobile host has to keep a history of the weighted average time that routes to a particular destination fluctuate until the route with the best metric is received (but not necessarily advertised). Of the paths with the same sequence number, those with the smallest metric will be used. By the natural way in which the routing tables are propagated, the sequence number is sent to all mobile computers, which may each decide to maintain a routing entry for that originating mobile computer.

Routes received in broadcasts are also advertised by the receiver when it subsequently broadcasts its routing information; the receiver adds an increment to the metric before advertising the route, because incoming packets will require one more hop to reach the destination (namely, the hop from the transmitter to the receiver). Again, we do not explicitly consider here the changes required to use metrics that do not use the hop count to the destination. One of the most important parameters to be chosen is the time between broadcasting the routing information packets. However, when any new or substantially modified route information is received by a mobile host, the new information will be retransmitted soon (subject to constraints imposed for damping route fluctuations), effecting the most rapid possible dissemination of routing information among all the cooperating mobile hosts. This quick rebroadcast introduces a new requirement for our protocols to converge as soon as possible. It would be calamitous if the movement of a mobile host caused a storm of broadcasts, degrading the availability of the wireless medium.

Mobile hosts cause broken links as they move from place to place. The broken link may be detected by the layer-2 protocol, or it may instead be inferred if no broadcasts have been received for a while from a former neighbor. A broken link is described by a metric of ∞ (i.e., any value greater than the maximum allowed metric). When a link to a next hop has broken, any route through that next hop is immediately assigned an ∞ metric and assigned an updated sequence number. Because this qualifies as a substantial route change, such modified routes are immediately disclosed in a broadcast routing information packet. Building information to describe broken links is the only situation when the sequence number is generated by any mobile host other than the destination mobile host. Sequence numbers defined by the originating mobile hosts are defined to be even numbers, and sequence numbers generated to indicate cm metrics are odd numbers. In this way, any "real" sequence numbers will supersede an m metric. When a node receives a co metric, and it has a later sequence number with a finite metric, it triggers a route update broadcast to disseminate the important news about that destination. In a very large population of mobile hosts, adjustments will likely be made in the time between broadcasts of the routing information packets. To reduce the amount of information carried in these packets, two types will be defined. One will carry all the available routing information; this is, as mentioned above, called a "full dump." The other type will carry only information changed since the last full dump; this is, as also mentioned, called an "incremental." By design, an incremental routing update should fit in one Network Protocol Data Unit (NPDU). The full dump will most likely require multiple NPDUs, even for relatively small populations of mobile hosts. Full dumps can be transmitted relatively infrequently when no movement of mobile hosts is occurring, When movement becomes frequent, and the size of an incremental approaches the size of an NPDU, then a full dump can be scheduled (so that the next incremental will be smaller). It is expected that mobile nodes will implement some means for determining which route changes are significant enough to be sent out with each incremental advertisement. For instance, when a stabilized route shows a different metric for some destination, that would likely constitute a significant change that needed to be advertised after stabilization. If a new sequence number for a route is received, but the metric stays the same, that would be unlikely to be considered as a significant change.

When a mobile host receives new routing information (usually in an incremental packet, as just described), that information is compared to the information already available from previous routing information packets. Any route with a more recent sequence number is used. Routes with older sequence numbers are discarded.

A route with a sequence number equal to an existing route is chosen if it has a "better" metric, and the existing route discarded, or stored as less preferable. The metrics for routes chosen from the newly received broadcast information are each incremented by one hop. Newly recorded routes are scheduled for immediate advertisement to the current mobile host's neighbors. Routes which show an improved metric are scheduled for advertisement at a time which depends on the average settling time for routes to the particular destination under consideration.

Timing skews between the various mobile hosts are expected. The broadcasts of routing information by the mobile hosts are to be regarded as somewhat asynchronous events, even though some regularity is expected. In such a population of independently transmitting agents, some fluctuation could develop using the above procedures for updating routes. It could turn out that a particular mobile host would receive new routing information in a pattern which causes it to consistently change routes from one next hop to another, even when the destination mobile host has not moved. This happens because there are two ways for new routes to be chosen; they might have a later sequence number, or predict how long to wait before advertising new routes.

3.4.2.2 Operating DSDV at Layer 2

The addresses stored in the routing tables will correspond to the layer at which this ad hoc networking protocol is operated. That is, operation at Layer 3 will use network-layer addresses for the next hop and destination addresses, and operation at Layer 2 will use Layer-2 MAC addresses. Using MAC addresses for the forwarding table does introduce a new requirement, however. The difficulty is that Layer-3 network protocols provide communication based on network addresses, and a way must be provided to resolve these Layer-3 addresses into MAC addresses. Otherwise, a multiplicity of different address resolution mechanisms would be put into place, and a corresponding loss of bandwidth in the wireless medium would be observed whenever the resolution mechanisms were utilized. This could be substantial because such mechanisms would require broadcasts and retransmitted broadcasts by every mobile host in the ad hoc network. Thus, unless special care is taken, every address resolution might look like a glitch in the normal operation of the network, which may well be noticeable to any active users. The solution proposed here, for operation at Layer 2, is to include Layer-3 protocol information along with the Layer-2 information. Each destination host would advertise which layer protocols it supports, and each mobile host advertising reachability to that destination would include, along with the advertisement, the information about the Layer-3 protocols supported at that destination. This information would only have to be transmitted when it changes, which occurs rarely. Changes would be transmitted as part of each incremental dump. Because each mobile host could support several Layer-3 protocols (and many will), this list would have to be variable in length.

3.4.2.3 Extending Base Station Coverage

Mobile computers will frequently be used in conjunction with base stations, which allow them to exchange data with other computers connected to the wired network. By participating in the DSDV protocol, base stations can extend their coverage beyond the range imposed by their wireless transmitters. When a base station participates in DSDV, it is shown as a default route in the tables transmitted by a mobile station. In this way, mobile stations within range of a base station can cooperate to effectively extend the range of the base station to serve other stations outside the range of the base station, as long as those other mobile stations are close to some other mobile station that is within range.

3.4.3 Optimized Link State Routing (OLSR) Protocol

OLSR is developed for mobile ad hoc networks. It operates as a table-driven, proactive protocol, that is, it exchanges topology information with other nodes of the network regularly. Each node selects a set of its neighbor nodes as "multipoint relays" (MPR). In OLSR, only nodes, selected as such MPRs, are responsible for forwarding control traffic, intended for diffusion into the entire network. MPRs provide an efficient mechanism for flooding control traffic by reducing the number of transmissions required. Nodes, selected as MPRs, also have a special responsibility when declaring link state information in the network. Indeed, the only requirement for OLSR to provide shortest path routes to all destinations is that MPR nodes declare link state information for their MPR selectors. Additional available link state information may be utilized, for example for redundancy.

Nodes which have been selected as multipoint relays by some neighbor node(s) announce this information periodically in their control messages. Thereby, a node announces to the network that it has reachability to the nodes which have selected it as an MPR. In route calculation, the MPRs are used to form the route from a given node to any destination in the network. Furthermore, the protocol uses the MPRs to facilitate efficient flooding of control messages in the network.

A node selects MPRs from among its one-hop neighbors with "symmetrical" (i.e., bidirectional) linkages. Therefore, selecting the route through MPRs automatically avoids the problems associated with data packet transfer over unidirectional links (such as the problem of not getting link-layer acknowledgments for data packets at each hop, for link layers employing this technique for unicast traffic). OLSR is developed to work independently from other protocols. Likewise, OLSR makes no assumptions about the underlying link layer. OLSR inherits the concept of forwarding and relaying from HIPERLAN (a MAC layer protocol), which is standardized by European Telecommunications Standards Institute (ETSI). The protocol is developed in the IPANEMA project (part of the Euclid program) and in the Perception Recognition Integration for Observation of Activity (PRIMA) project (part of the RNRT program).

OLSR is a proactive routing protocol for mobile ad hoc networks. It is well suited to large and dense mobile networks, as the optimization achieved using the MPRs works well in this context. The larger and more dense a network, the more optimization can be achieved as compared to the classic link state algorithm. OLSR uses hop-by-hop routing, that is, each node uses its local information to route packets. OLSR is well suited for networks, where the traffic is random and sporadic between a larger set of nodes rather than being almost exclusively between a small specific set of nodes. As a proactive protocol, OLSR is also suitable for scenarios where the communicating pairs change over time: no additional control traffic is generated in this situation because routes are maintained for all known destinations at all times.

3.4.3.1 Protocol Overview

OLSR is, as discussed in the previous subsection, a proactive routing protocol for mobile ad hoc networks. The protocol inherits the stability of a link state algorithm and has the advantage of having routes immediately available when needed due to its proactive nature. OLSR is an optimization over the classical link state protocol, tailored for mobile ad hoc networks. OLSR minimizes the overhead from flooding of control traffic by using only selected nodes, called MPRs, to retransmit control messages. This technique significantly reduces the number of retransmissions required to flood a message to all nodes in the network. Secondly, OLSR requires only a partial link state to be flooded to provide shortest path routes. The minimal set of link state information required is that all nodes selected as MPRs must declare the links to their MPR selectors. Additional topological information, if present, *may* be utilized, for example for redundancy purposes.

OLSR may optimize the reactivity to topological changes by reducing the maximum time interval for periodic control message transmission. Furthermore, as OLSR continuously maintains routes to all destinations in the network, the protocol is beneficial for traffic patterns where a large subset of nodes are communicating with another large subset of nodes, and where the source–destination pairs are changing over time. The protocol is particularly suited for large and dense networks, as the optimization done using MPRs works well in this context. The larger and more dense a network, the more optimization can be achieved as compared to the classic link state algorithm.

OLSR is designed to work in a completely distributed manner and does not depend on any central entity. The protocol does not require reliable transmission of control messages: each node sends control messages periodically, and can therefore sustain a reasonable loss of some such messages. Such losses occur frequently in radio networks due to collisions or other transmission problems. Also, OLSR does not require sequenced delivery of messages. Each control message contains a sequence number, which is incremented for each message. Thus the recipient of a control message can, if required, easily identify which information is more recent—even if messages have been reordered while in transmission. Furthermore, OLSR provides support for protocol extensions such as sleep mode operation and multicast routing. Such extensions may be introduced as additions to the protocol without breaking backwards compatibility with earlier versions. OLSR does not require any changes to the format of Internet Protocol (IP) packets. Thus any existing IP stack can be used as is; the protocol only interacts with routing table management.

3.4.3.2 Multipoint Relays (MPRs)

The idea of multipoint relays is to minimize the overhead of flooding messages in the network by reducing redundant retransmissions in the same region. Each node in the network selects a set of nodes in its symmetric one-hop neighborhood, which may retransmit its messages. This set of selected neighbor nodes is called the MPR set of that node. The neighbors of node N which are *not* in its MPR set receive and process broadcast messages but do not retransmit broadcast messages received from node N.

Each node selects its MPR set from among its one-hop symmetric neighbors. This set is selected such that it covers (in terms of radio range) all symmetric strict two-hop nodes. The MPR set of *N*, denoted as MPR (*N*), is then an arbitrary subset of the symmetric one-hop neighborhood of *N* which satisfies the following condition: every node in the symmetric strict two-hop neighborhood of *N* must have a symmetric link toward MPR (*N*). The smaller an MPR set, the less control traffic overhead results from the routing protocol. Each node maintains information about the set of neighbors that have selected it as an MPR. This set is called the "MPR selector set" of a node. A node obtains this information from periodic Hello messages received from the neighbors. Upon receipt of this MPR selector information, each node calculates and updates its route to each destination. Therefore, the route is a sequence of hops through the multipoint relays from source to destination.

3.4.3.3 Protocol Functioning

OLSR is modularized into a "core" of functionality, which is always required for the protocol to operate, and a set of auxiliary functions. The core specifies, in its own right, a protocol able to provide routing in a stand-alone MANET. Each auxiliary function provides additional functionality, which may be applicable in specific scenarios (e.g., in case a node is providing connectivity between the MANET and another routing domain). All auxiliary functions are compatible, to the extent where any (sub-)set of auxiliary functions may be implemented with the core. Furthermore, the protocol allows heterogeneous nodes—that is, nodes which implement different subsets of the auxiliary functions—to coexist in the network. The purpose of dividing the functioning of OLSR into core functionality and a set of auxiliary functions is to provide a simple and easy-to-comprehend protocol, and to provide a way of only adding complexity where specific additional functionality is required.

3.4.3.4 Core Functioning

The core functionality of OLSR specifies the behavior of a node, equipped with OLSR interfaces participating in the MANET and running OLSR as a routing protocol. This includes a universal specification of OLSR protocol messages and their transmission through the network, as well as link sensing, topology diffusion, and route calculation.

Specifically, the core is made up from the following components.

3.4.2.1.1 Packet Format and Forwarding

A universal specification of the packet format and an optimized flooding mechanism serves as the transport mechanism for all OLSR control traffic.

3.4.2.1.2 Link Sensing

Link sensing is accomplished through periodic emission of Hello messages over the interfaces through which connectivity is checked. A separate Hello message is generated for each interface. Resulting from link sensing is a local link set describing links between "local interfaces" and "remote interfaces," that is, interfaces on neighbor nodes. If sufficient information is provided by the link layer, this may be utilized to populate the local link set instead of a Hello message exchange.

3.4.2.1.3 Neighbor Detection

Given a network with only single interface nodes, a node may deduct the neighbor set directly from the information exchanged as part of link sensing: the "main address" of a single interface node is, by definition, the address of the only interface on that node. In a network with multiple interface nodes, additional information is required to map interface addresses to main addresses (and, thereby, to nodes). This additional information is acquired through Multiple Interface Declaration (MID) messages.

3.4.2.1.4 MPR Selection and MPR Signaling

The objective of MPR selection is for a node to select a subset of its neighbors such that a broadcast message, retransmitted by these selected neighbors, will be received by all nodes two hops away. The MPR set of a node is computed such that it, for each interface, satisfies this condition. The information required to perform this calculation is acquired through the periodic exchange of Hello messages.

3.4.2.1.5 Topology Control Message Diffusion

Topology control messages are diffused with the purpose of providing each node in the network with sufficient link state information to allow route calculation.

3.4.2.1.6 Route Calculation

Given the link state information acquired through periodic message exchange, as well as the interface configuration of the nodes, the routing table for each node can be computed.

3.4.4 Fisheye State Routing (FSR)

FSR is an implicit hierarchical routing protocol. It uses the "fisheye" technique proposed by Kleinrock and Stevens, where the technique was used to reduce the size of information required to represent graphical data. The eye of a fish captures with high detail the pixels near the focal point. The detail decreases as the distance from the focal point increases. In routing, the fisheye approach translates to maintaining accurate distance and path quality information about the immediate neighborhood of a node, with progressively less detail as the distance increases.

FSR is functionally similar to LSR in that it maintains a topology map at each node. The key difference is the way in which routing information is disseminated. In LSR, link state packets are generated and flooded into the network whenever a node detects a topology change. In FSR, link state packets are not flooded. Instead, nodes maintain a link state table based on the up-to-date information received from neighboring nodes, and periodically exchange it with their local neighbors only (no flooding). Through this exchange process, the table entries with larger sequence numbers replace the ones with smaller sequence numbers. The FSR periodic table exchange resembles the vector exchange in DBF (or, more precisely, DSDV, where the distances are updated according to the time stamp or sequence number assigned by the node originating the update). However, in FSR link states rather than distance vectors are propagated. Moreover, like in Linked State Routing (LSR), a full topology map is kept at each node and shortest paths are computed using this map.

In a wireless environment, a radio link between mobile nodes may experience frequent disconnects and reconnects. The LSR Protocol releases a link state update for each such change, which floods the network and causes excessive overhead. FSR avoids this problem by using periodic, instead of event-driven, exchange of the topology map, greatly reducing the control message overhead. When network size grows large, the update message could consume a considerable amount of bandwidth, which depends on the update period. To reduce the size of update messages without seriously affecting routing accuracy, FSR uses the fisheye technique. Figure 3.1 illustrates the application of fisheye in a mobile, wireless network.

The circles with different shades of gray define the fisheye scopes with respect to the center node (node 11). The scope is defined as the set of nodes that can be reached within a given number of hops. In our case, three scopes are shown for 1, 2, and > 2 hops, respectively. Nodes are color coded as black, gray, and white accordingly. The number of levels and the radius of each scope will depend on the size of the network. The reduction of routing update overhead is obtained by using different exchange periods for different entries in the routing table.

More precisely, entries corresponding to nodes within the smaller scope are propagated to the neighbors with the highest frequency. Referring to Figure 3.2, entries in bold are exchanged most frequently. The rest of the entries are sent out at a lower frequency. As a result, a considerable fraction of link state entries are sup-

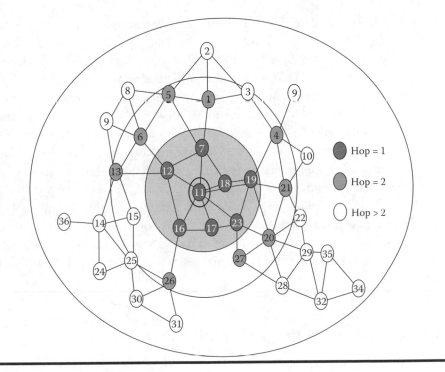

Figure 3.1 Scope of fisheye.

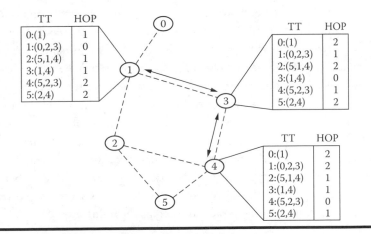

Figure 3.2 Message reduction using fisheye.

pressed in a typical update, thus reducing the message size. This strategy produces timely updates from nearby stations, but creates large latencies from stations afar.

However, the imprecise knowledge of the best path to a distant destination is compensated by the fact that the route becomes progressively more accurate as the packet gets closer to the destination. As the network size grows large, a "graded" frequency update plan must be used across multiple scopes to keep the overhead low.

The FSR concept originates from the Global State Routing (GSR) Protocol. GSR can be viewed as a special case of FSR, in which there is only one fisheye-scope level. As a result, the entire topology table is exchanged among neighbors. Clearly, this consumes a considerable amount of bandwidth when the network size becomes large. Through updating link state information with different frequencies depending on the scope distance, FSR scales well to large network size and keeps overhead low without compromising route computation accuracy when the destination is near. By retaining a routing entry for each destination, FSR avoids the extra work of "finding" the destination (as in on-demand routing) and thus maintains low single-packet transmission latency. As mobility increases, routes to remote destinations become less accurate. However, when a packet approaches its destination, it finds increasingly accurate routing instructions as it enters sectors with a higher refresh rate.

3.5 Reactive Routing Protocols

Reactive routing protocols for mobile ad hoc networks are also, as discussed previously, called "on-demand" routing protocols. In a reactive routing protocol, routing paths are searched only when needed. A route discovery operation invokes a route determination procedure. The discovery procedure terminates when either a route has been found or no route is available after examination for all route permutations. In a mobile ad hoc network, active routes may be disconnected due to node mobility. Therefore, route maintenance is an important operation of reactive routing protocols. Compared to the proactive routing protocols for mobile ad hoc networks, less control overhead is a distinct advantage of the reactive routing protocols. Thus, reactive routing protocols have better scalability than proactive routing protocols in mobile ad hoc networks. However, when using reactive routing protocols, source nodes may suffer from long delays for route searching before they can forward data packets. In this section, we will discuss several typical reactive mobile ad hoc network routing protocols, such as the following:

- AODV
- DSR
- Temporally Ordered Routing Algorithm (TORA)
- CBRP
- LAR
- Ant-Colony-Based Routing Algorithm (ARA)

3.5.1 Ad Hoc On-Demand Distance Vector (AODV)

An ad hoc network is the cooperative engagement of a collection of mobile nodes without the required intervention of any centralized access point or existing infrastructure. AODV is a novel algorithm for the operation of such ad hoc networks. Each mobile host operates as a specialized router, and routes are obtained as needed (i.e., on demand with little or no reliance on periodic advertisements). The AODV routing algorithm is quite suitable for a dynamic self-starting network as required by users wishing to utilize ad hoc networks. AODV provides loop-free routes even while repairing broken links. Because the protocol does not require global periodic routing advertisements, the demand on the overall bandwidth available to the mobile nodes is substantially less than in those protocols that do necessitate such advertisements.

AODV uses symmetric links between neighboring nodes. It does not attempt to follow paths between nodes when one of the nodes cannot hear the other one. Nodes do not lie on active paths; they neither maintain any routing information nor participate in any periodic routing table exchanges. Further, a node does not have to discover and maintain a route to another node until the two need to communicate unless the former node is offering its services as an intermediate forwarding station to maintain connectivity between two other nodes. When the local connectivity of the mobile node is of interest, each mobile node can become aware of the other nodes in its neighborhood by the use of several techniques, including local (not systemwide) broadcasts known as Hello messages. The routing tables of the nodes within the neighborhood are organized to optimize response time to local movements and provide quick response time for requests for establishment of new routes.

The algorithm's primary objectives are as follows:

- To broadcast discovery packets only when necessary.
- To distinguish between local connectivity management neighborhood detection and general topology maintenance.
- To disseminate information about changes in local connectivity to those neighboring mobile nodes that are likely to need the information.

AODV uses a broadcast route discovery mechanism as is also used with modifications in the DSR algorithm. Instead of source routing, however, AODV relies on dynamically establishing route table entries at intermediate nodes. This difference pays off in networks with many nodes where a larger overhead is incurred by carrying source routes in each data packet.

To maintain the most recent routing information between nodes, we borrow the concept of destination sequence numbers from DSDV. Unlike in DSDV, however, each ad hoc node maintains a monotonically increasing sequence number counter which is used to supersede stale cached routes. The combination of these techniques yields an algorithm that uses bandwidth efficiently by minimizing the network load for control, and data traffic is responsive to changes in topology and ensures loop-free routing.

3.5.1.1 Path Discovery

The path discovery process is initiated whenever a source node needs to communicate with another node for which it has no routing information in its table. Every node maintains two separate counters: a node sequence number and a broadcast ID. The source node initiates path discovery by broadcasting a Route REQuest (RREQ) packet to its neighbors.

The RREQ contains the following fields:

<source_addr source sequence# broadcast id dest_addr dest sequence# hop cnt>

The pair *<source_addr, broadcast_id>* uniquely identifies an RREQ. *broadcast_id* is incremented whenever the source issues a new RREQ. Each neighbor either satisfies the RREQ by sending a Route REPly (RREP) back to the source, or broadcasts the RREQ to its own neighbors after increasing the *hop_cnt*. Notice that a node may receive multiple copies of the same route broadcast packet from various neighbors. When an intermediate node receives an RREQ, if it has already received an RREQ with the same *broadcast_id* and source address, it drops the redundant RREQ and does not rebroadcast it. If a node cannot satisfy the RREQ, it keeps track of the following information to implement the reverse-path setup as well as the forward-path setup that will accompany the transmission of the eventual RREP.

- Destination IP address
- Source IP address
- Broadcast ID
- Expiration time for reverse-path route entry
- Source node's sequence number

3.5.1.2 Reverse-Path Setup

There are two sequence numbers (in addition to the *broadcast_id*) included in an RREQ: the source sequence number and the last destination sequence number known to the source. The source sequence number is used to maintain freshness information about the reverse route to the source, and the destination sequence number specifies how fresh a route to the destination must be before it can be accepted by the source. As the RREQ travels from a source to various destinations, it automatically sets up the reverse path from all nodes back to the source. To set up a reverse path, a node records the address of the neighbor from which it received the first copy of the RREQ. These reverse-path route entries are maintained for at least enough time for the RREQ to traverse the network and produce a reply to the sender.

3.5.1.3 Forward-Path Setup

Eventually, an RREQ will arrive at a node (possibly the destination itself) that possesses a current route to the destination. The receiving node first checks that the RREQ was received over a bidirectional link. If an intermediate node has a route entry for the desired destination, it determines whether the route is current by comparing the destination sequence number in its own route entry to the destination sequence number in the RREQ. If the RREQ's sequence number for the destination is greater than that recorded by the intermediate node, the intermediate node must not use its recorded route to respond to the RREQ. Instead, the intermediate node rebroadcasts the RREQ. The intermediate node can reply only when it has a route with a sequence number that is greater than or equal to that contained in the RREQ. If it does have a current route to the destination and if the RREQ has not been processed previously, the node then unicasts a route reply packet (RREP) back to its neighbor from which it received the RREQ.

An RREP contains the following information:

<source_addr, dest_addr, dest_sequence #, hop_cnt, lifetime>

By the time a broadcast packet arrives at a node that can supply a route to the destination, a reverse path has been established to the source of the RREQ. As the RREP travels back to the source, each node along the path sets up a forward pointer to the node from which the RREP came, updates its timeout information for route entries to the source and destination, and records the latest destination sequence number for the requested destination. The forward path setup as the RREP travels from the destination D to the source node S. Nodes that are not along the path determined by the RREP will time out after ACTIVE_ROUTE_TIMEOUT (3000 milliseconds) and will delete the reverse pointers.

A node receiving an RREP propagates the first RREP for a given source node toward that source. If it receives further RREPs, it updates its routing information and propagates the RREP only if the RREP contains either a greater destination sequence number than the previous RREP or the same destination sequence number with a smaller hop count. It suppresses all other RREPs it receives. This decreases the number of RREPs propagating toward the source while also ensuring the quickest and most up-to-date routing information. The source node can begin data transmission as soon as the first RREP is received and can later update its routing information if it learns of a better route.

3.5.1.4 Route Table Management

In addition to the source and destination sequence numbers, other useful information is also stored in the route table entries and is called the "soft state" associated with the entry. Associated with reverse-path routing entries is a timer called the "route request expiration timer." The purpose of this timer is to purge reverse-path routing entries from those nodes that do not lie on the path from the source to the destination. The expiration time depends upon the size of the ad hoc network. Another important parameter associated with routing entries is the route-caching timeout, or the time after which the route is considered to be invalid. In each routing table entry, the address of active neighbors through which packets for the given destination are received is also maintained. A neighbor is considered active for that destination if it originates or relays at least one packet for that destination within the most recent active timeout period. This information is maintained so that all active source nodes can be notified when a link along a path to the destination breaks. A route entry is considered active if it is in use by any active neighbors. The path from a source to a destination, which is followed by packets along active route entries, is called an "active path." Note that, as with DSDV, all routes in the route table are tagged with destination sequence numbers, which guarantee that no routing loops can form, even under extreme conditions of out-of-order packet delivery and high node mobility. A mobile node maintains a route table entry for each destination of interest. Each route table entry contains the following information:

- Destination
- Next hop
- Number of hops (metric)
- Sequence number for the destination
- Active neighbors for this route
- Expiration time for the route table entry

Each time a route entry is used to transmit data from a source toward a destination, the timeout for the entry is reset to the current time plus the active route timeout. If a new route is offered to a mobile node, the mobile node compares the destination sequence number of the new route to the destination sequence number for the current route. The route with the greater sequence number is chosen. If the sequence numbers are the same, then the new route is selected only if it has a smaller metric (a fewer numbers of hops) to the destination.

3.5.1.5 Path Maintenance

Movement of nodes not lying along an active path does not affect the routing to that path's destination. If the source node moves during an active session, it can reinitiate the route discovery procedure to establish a new route to the destination.

When either the destination or some intermediate node moves, a special RREP is sent to the affected source nodes. Periodic Hello messages can be used to ensure symmetric links, as well as to detect link failures. Alternatively, and with far less latency, such failures could be detected by using Link-Layer ACKnowledgments (LLACKs). A link failure is also indicated if attempts to forward a packet to the next hop fail.

Once the next hop becomes unreachable, the node upstream of the break propagates an unsolicited RREP with a fresh sequence number, that is, a sequence number that is one greater than the previously known sequence number and hop count of all active upstream neighbors. Those nodes subsequently relay that message to their active neighbors and so on. This process continues until all active source nodes are notified; it terminates because AODV maintains only loop-free routes, and there are only a finite number of nodes in the ad hoc network.

Upon receiving notification of a broken link, source nodes can restart the discovery process if they still require a route to the destination. To determine whether a route is still needed, a node may check whether the route has been used recently as well as inspect upper-level protocol control blocks to see whether connections remain open using the indicated destination. If the source node or any other node along the previous route decides it would like to rebuild the route to the destination, it sends out an RREQ with a destination sequence number of one greater than the previously known sequence number to ensure that it builds a new viable route, and that no nodes reply if they still regard the previous route as valid.

3.5.1.6 Local Connectivity Management

Nodes learn of their neighbors in one of two ways. Whenever a node receives a broadcast from a neighbor, it updates its local connectivity information to ensure that it includes this neighbor. In the event that a node has not sent any packets to all of its active downstream neighbors within a Hello interval, it broadcasts to its neighbors a Hello message, a special unsolicited RREP containing its identity and sequence number. The node's sequence number is not changed for Hello message transmissions. This Hello message is prevented from being rebroadcast outside the neighborhood of the node because it contains a Time-To-Live (TTL) value of one. Neighbors that receive this packet update their local connectivity information to the node. Receiving a broadcast or a Hello message from a new neighbor or failing to receive allowed Hello loss consecutive Hello messages from a node previously in the neighborhood is an indication that the local connectivity has changed. Failing to receive Hello messages from inactive neighbors does not trigger any protocol action. If Hello messages are not received from the next hop along an active path, the active neighbors using that next hop are sent notification of link failure. We have determined that the optimal value for allowed Hello loss is two. The local connectivity management with Hello messages can also be used to ensure that only nodes with bidirectional connectivity are considered to be neighbors. For this purpose, each Hello sent by a node lists the nodes from which it has heard. Each node checks to make sure that it uses only routes to neighbors that have heard the node's Hello message. To save local bandwidth, such checking should be performed only if explicitly configured into the nodes.

3.5.2 Dynamic Source Routing (DSR) Protocol

The DSR Protocol is a simple and efficient routing protocol designed specifically for use in multihop wireless ad hoc networks of mobile nodes. Using DSR, the network is completely self-organizing and self-configuring, requiring no existing network infrastructure or administration. Network nodes (computers) cooperate to forward packets for each other to allow communication over multiple "hops" between nodes not directly within wireless transmission range of one another. As nodes in the network move about or join or leave the network, and as wireless transmission conditions such as sources of interference change, all routing is automatically determined and maintained by the DSR Routing Protocol. Because the number or sequence of intermediate hops needed to reach any destination may change at any time, the resulting network topology may be quite rich and rapidly changing.

The DSR Protocol allows nodes to dynamically discover a source route across multiple network hops to any destination in the ad hoc network. Each data packet sent then carries in its header the complete, ordered list of nodes through which the packet must pass, allowing packet routing to be trivially loop-free and avoiding the need for up-to-date routing information in the intermediate nodes through which the packet is forwarded. By including this source route in the header of each data packet, other nodes forwarding or overhearing any of these packets may also easily cache this routing information for future use.

While designing DSR, we needed to create a routing protocol that had very low overhead yet was able to react quickly to changes in the network, providing highly reactive service to help ensure successful delivery of data packets in spite of node movement or other changes in network conditions.

3.5.2.1 Overview and Important Properties of the Protocol

The DSR Protocol is composed of two mechanisms that work together to allow the discovery and maintenance of source routes in the ad hoc network:

- *Route discovery* is the mechanism by which a node S wishing to send a packet to a destination node D obtains a source route to D. Route discovery is used only when S attempts to send a packet to D and does not already know a route to D.
- *Route maintenance* is the mechanism by which node S is able to detect, while using a source route to D, if the network topology has changed such that it can

no longer use its route to D because a link along the route no longer works. When route maintenance indicates a source route is broken, S can attempt to use any other route it happens to know to D, or can invoke route discovery again to find a new route. Route maintenance is used only when S is actually sending packets to D. Route discovery and route maintenance each operate entirely *on demand*. In particular, unlike other protocols, DSR requires *no* periodic packets of *any kind* at *any level* within the network. For example, DSR does not use any periodic routing advertisement, link status sensing, or neighbor detection packets, and does not rely on these functions from any underlying protocols in the network. This entirely on-demand behavior and lack of periodic activity allow the number of overhead packets caused by DSR to scale all the way down to *zero*, when all nodes are approximately stationary with respect to each other and all routes needed for current communication have already been discovered. As nodes begin to move more or as communication patterns change, the routing packet overhead of DSR *automatically* scales to only that needed to track the routes currently in use.

In response to a single route discovery (as well as through routing information from other packets overheard), a node may learn and cache multiple routes to any destination. This allows the reaction to routing changes to be much more rapid, because a node with multiple routes to a destination can try another cached route if the one it has been using should fail. This caching of multiple routes also avoids the overhead of needing to perform a new route discovery each time a route in use breaks.

The operations of route discovery and route maintenance in DSR are designed to allow unidirectional links and asymmetric routes to be easily supported. In particular, as noted in Section 3.2 (above), in wireless networks, it is possible that a link between two nodes may not work equally well in both directions, due to differing antenna or propagation patterns or sources of interference. DSR allows such unidirectional links to be used when necessary, improving overall performance and network connectivity in the system.

DSR also supports internetworking between different types of wireless networks, allowing a source route to be composed of hops over a combination of any types of networks available. For example, some nodes in the ad hoc network may have only short-range radios, while other nodes have both short-range and long-range radios; the combination of these nodes can be considered by DSR as a single ad hoc network. In addition, the routing of DSR has been integrated into standard Internet routing, where a "gateway" node connected to the Internet also participates in the ad hoc network routing protocols, and has been integrated into Mobile IP routing, where such a gateway node also serves the role of a Mobile IP foreign agent.

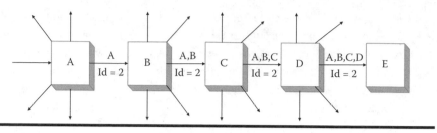

Figure 3.3   Route discovery example: node A is the initiator, and node E is the target.

3.5.2.2 Basic DSR Route Discovery

When some node S originates a new packet destined to some other node D, it places in the header of the packet a "source route" giving the sequence of hops that the packet should follow on its way to D. Normally, S will obtain a suitable source route by searching its "route cache" of routes previously learned, but if no route is found in its cache, it will initiate the route discovery protocol to dynamically find a new route to D.

In this case, we call S the "initiator" and D the "target" of the route discovery. For example, Figure 3.3 illustrates an example of route discovery, in which a node A is attempting to discover a route to node E. To initiate the route discovery, A transmits a route request message as a single local broadcast packet, which is received by (approximately) all nodes currently within wireless transmission range of A. Each route request message identifies the initiator and target of the route discovery, and also contains a unique "request ID," determined by the initiator of the request. Each route request also contains a record listing the address of each intermediate node through which this particular copy of the route request message has been forwarded. This route record is initialized to an empty list by the initiator of the route discovery.

When another node receives a route request, if it is the target of the route discovery, it returns a route reply message to the initiator of the route discovery, giving a copy of the accumulated route record from the route request; when the initiator receives this route reply, it caches this route in its route cache for use in sending subsequent packets to this destination. Otherwise, if this node receiving the route request has recently seen another route request message from this initiator bearing this same request ID, or if it finds that its own address is already listed in the route record in the route request message, it discards the request. Otherwise, this node appends its own address to the route record in the route request message and propagates it by transmitting it as a local broadcast packet (with the same request ID).

In returning the route reply to the initiator of the route discovery, such as node E replying back to A in Figure 3.3, node E will typically examine its own route cache for a route back to A and, if found, will use it for the source route for delivery of the packet containing the route reply. Otherwise, E may perform its own route discovery for target node A, but to avoid possible infinite recursion of route discoveries, it must piggyback this route reply on its own route request message for A. It is also possible to piggyback other small data packets, such as a Transmission Control Protocol (TCP) Synchronization (SYN) packet, on a route request using this same mechanism. Node E could also simply reverse the sequence of hops in the route record that it is trying to send in the route reply, and use this as the source route on the packet carrying the route reply itself.

For MAC protocols such as the Institute of Electrical and Electronics Engineers (IEEE) 802.11 Protocol that require a bidirectional frame exchange as part of the MAC protocol, this route reversal is preferred as it avoids the overhead of a possible second route discovery, and it tests the discovered route to ensure it is bidirectional before the route discovery initiator begins using the route. However, this technique will prevent the discovery of routes using unidirectional links. In wireless environments where the use of unidirectional links is permitted, such routes may in some cases be more efficient than those with only bidirectional links, or they may be the only way to achieve connectivity to the target node.

When initiating a route discovery, the sending node saves a copy of the original packet in a local buffer called the "send buffer." The send buffer contains a copy of each packet that cannot be transmitted by this node because it does not yet have a source route to the packet's destination. Each packet in the send buffer is stamped with the time that it was placed into the buffer and is discarded after residing in the send buffer for some timeout period; if necessary for preventing the send buffer from overflowing, a FIFO or other replacement strategy can also be used to evict packets before they expire.

While a packet remains in the send buffer, the node should occasionally initiate a new route discovery for the packet's destination address. However, the node must limit the rate at which such new route discoveries for the same address are initiated, because it is possible that the destination node is not currently reachable. In particular, due to the limited wireless transmission range and the movement of the nodes in the network, the network may at times become partitioned, meaning that there is currently no sequence of nodes through which a packet could be forwarded to reach the destination. Depending on the movement pattern and the density of nodes in the network, such network partitions may be rare or may be common.

If a new route discovery was initiated for each packet sent by a node in such a situation, a large number of unproductive route request packets would be propagated throughout the subset of the ad hoc network reachable from this node. To reduce the overhead from such route discoveries, we use exponential backoff to limit the rate at which new route discoveries may be initiated by any node for the same target. If the node attempts to send additional data packets to this same node more frequently than this limit, the subsequent packets should be buffered in the send buffer until a route reply is received, but the node must not initiate a new route discovery until the minimum allowable interval between new route discoveries for this target has been reached. This limitation on the maximum rate

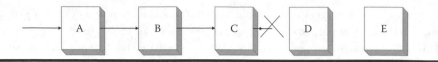

Figure 3.4 Route maintenance example: node C is unable to forward a packet from A to E over its link to next-hop D.

of route discoveries for the same target is similar to the mechanism required by Internet nodes to limit the rate at which ad hoc routing protocol requests are sent for any single target IP address.

3.5.2.3 Basic DSR Route Maintenance

When originating or forwarding a packet using a source route, each node transmitting the packet is responsible for confirming that the packet has been received by the next hop along the source route; the packet is retransmitted (up to a maximum number of attempts) until this confirmation of receipt is received. For example, in the situation illustrated in Figure 3.4, node A has originated a packet for E using a source route through intermediate nodes B, C, and D. In this case, node A is responsible for receipt of the packet at B, node B is responsible for receipt at C, node C is responsible for receipt at D, and, finally, node D is responsible for receipt at the destination E.

This confirmation of receipt in many cases may be provided at no cost to DSR, either as an existing standard part of the MAC protocol in use (such as the link-level acknowledgment frame defined by IEEE 802.11) or by a "passive acknowledgment" (in which B confirms receipt at C by overhearing C transmit the packet to forward it on to D). If neither of these confirmation mechanisms is available, the node transmitting the packet may set a bit in the packet's header to request that a DSR-specific software acknowledgment be returned by the next hop; this software acknowledgment will normally be transmitted directly to the sending node, but if the link between these two nodes is unidirectional, this software acknowledgment may travel over a different multihop path.

If the packet is retransmitted by some hop the maximum number of times and no receipt confirmation is received, this node returns a route error message to the original sender of the packet, identifying the link over which the packet could not be forwarded. For example, in Figure 3.4, if C is unable to deliver the packet to the next hop D, then C returns a route error to A, stating that the link from C to D is currently "broken."

Node A then removes this broken link from its cache; any retransmission of the original packet is a function for upper-layer protocols such as TCP. For sending such a retransmission or other packets to this same destination E, if A has in its route cache another route to E (for example, from additional route replies from its earlier route discovery, or from having overheard sufficient routing information from other packets), it can send the packet using the new route immediately. Otherwise, it may perform a new route discovery for this target.

3.5.3 Temporally Ordered Routing Algorithm (TORA)

TORA is a distributed routing protocol for mobile, multihop wireless networks. Its intended use is for the routing of IP datagrams within an autonomous system. The basic, underlying algorithm is neither a distance vector nor a link state; it is one of a family of algorithms referred to as "link-reversal" algorithms. The protocol's reaction is structured as a temporally ordered sequence of diffusing computations, each computation consisting of a sequence of directed link reversals. The protocol is highly adaptive, efficient, and scalable, and is well suited for use in large, dense, mobile networks. In these networks, the protocol's reaction to link failures typically involves only a localized "single pass" of the distributed algorithm. This desirable behavior is achieved through the use of a physical or logical clock to establish the "temporal order" of topological change events. The established temporal ordering is subsequently used to structure (or order) the algorithm's reaction to topological changes.

TORA's design is predicated on the notion that a routing algorithm that is well suited for operation in this environment should possess the following properties:

- Executes distributedly.
- Provides loop-free routes.
- Provides multiple routes (i.e., to reduce the frequency of reactions to topological changes, and potentially to alleviate congestion).
- Establishes routes quickly (i.e., so they may be used before the topology changes).
- Minimizes communication overhead by localizing algorithmic reaction to topological changes when possible (i.e., to conserve available bandwidth and increase scalability).

Routing optimality (i.e., determination of the shortest path) is of less importance. It is also not necessary (or desirable) to maintain routes between every source–destination pair at all times. The overhead expended to establish a route between a given source–destination pair will be wasted if the source does not require the route prior to its invalidation due to topological changes.

TORA is designed to minimize reaction to topological changes. A key concept in its design is that it decouples the generation of potentially far-reaching control message propagation from the rate of topological changes. Control messaging is typically localized to a very small set of nodes near the change without having to resort to a dynamic, hierarchical routing solution with its attendant complexity. TORA includes a secondary mechanism, which allows far-reaching control message propagation as a means of infrequent route optimization and soft-state route verification. This propagation occurs periodically at a very low rate and is independent of the network topology dynamics.

TORA is distributed in that nodes need only to maintain information about adjacent nodes (i.e., one-hop knowledge). It guarantees all routes are loop free, and typically provides multiple routes for any source–destination pair that requires a route. TORA is "source initiated" and quickly creates a set of routes to a given destination only when desired. Because multiple routes are typically established and having a single route is sufficient, many topological changes require no reaction at all. Following topological changes that do require reaction, the protocol quickly reestablishes valid routes. This ability to initiate and react infrequently serves to minimize communication overhead. Finally, in the event of a network partition, the protocol detects the partition and erases all invalid routes.

A logically separate version of TORA is run for each destination to which routing is required. The following discussion focuses on a single version running for a given destination, j. TORA can be separated into three basic functions: creating routes, maintaining routes, and erasing routes. Creating a route from a given node to the destination requires establishment of a sequence of directed links leading from the node to the destination. This function is only initiated when a node with no directed links requires a route to the destination. Thus, creating routes essentially corresponds to assigning directions to links in an undirected network or portion of the network. The method used to accomplish this is an adaptation of the query–reply process, which builds a Directed Acyclic Graph (DAG) rooted at the destination (i.e., the destination is the only node with no downstream links). Such a DAG will be referred to as a "destination-oriented DAG." "Maintaining routes" refers to reacting to topological changes in the network in a manner such that routes to the destination are reestablished within a finite time—meaning that its directed portions return to a destination-oriented DAG within a finite time. However, the Gafni–Bertsekas (GB) algorithms are designed for operation in connected networks. Due to instability exhibited by these algorithms in portions of the network that become partitioned from the destination, they are deemed unacceptable for the current task. TORA incorporates a new algorithm, in the same general class, that is more efficient in reacting to topological changes and capable of detecting a network partition. This leads to the third function, erasing routes. Upon detection of a network partition, all links (in the portion of the network that has become partitioned from the destination) must be marked as undirected to erase invalid routes.

TORA accomplishes these three functions through the use of three distinct control packets: query (QRY), update (UPD), and clear (CLR). QRY packets are used for creating routes, UPD packets are used for both creating and maintaining routes, and CLR packets are used for erasing routes.

3.5.4 Cluster-Based Routing Protocol (CBRP)

CBRP is a routing protocol designed for use in mobile ad hoc networks. The protocol divides the nodes of the ad hoc network into a number of overlapping or disjoint two-hop-diameter clusters in a distributed manner. A clusterhead is elected for each cluster to maintain cluster membership information. Intercluster routes are discovered dynamically using the cluster membership information kept at each clusterhead. By clustering nodes into groups, the protocol efficiently minimizes the flooding traffic during route discovery and speeds up this process as well. Furthermore, the protocol takes into consideration the existence of unidirectional links and uses these links for both intracluster and intercluster routing.

The two major new features that have been added to the protocol are route shortening and local repair. Both features make use of the two-hop-topology information maintained by each node through the broadcasting of Hello messages. The route-shortening mechanism dynamically shortens the source route of the data packet being forwarded and informs the source about the better route. Local route repair patches a broken source route automatically and avoids route rediscovery by the source.

There are several major difficulties for designing a routing protocol for a MANET. Firstly and most importantly, a MANET has a dynamically changing topology due to the movement of mobile nodes, which favors routing protocols that dynamically discover routes over conventional distance-vector routing protocols. Secondly, the fact that a MANET lacks any structure makes IP subnetting inefficient. However, routing protocols that are flat (i.e., have no hierarchy) might suffer from excessive overhead when scaled up. Third, links in mobile networks could be asymmetric at times. If a routing protocol relies only on bidirectional links, the size and connectivity of the network may be severely limited; in other words, a protocol that makes use of unidirectional links can significantly reduce network partitions and improve routing performance.

CBRP has the following features:

- Fully distributed operation.
- Less flooding traffic during the dynamic route discovery process.
- Explicit exploitation of unidirectional links that would otherwise be unused.
- Broken routes could be repaired locally without rediscovery.
- Suboptimal routes could be shortened as they are used.

In these protocols, clusters are introduced to minimize updating overhead during topology change. However, the overhead for maintaining up-to-date information about the whole network's cluster membership and intercluster routing information at each and every node to route a packet is considerable. As network topology changes from time to time due to node movement, the effort to maintain such up-to- date information is expensive and rarely justified as such global cluster membership information is obsolete long before it is used. In comparison, simpler and smaller clusters are used; however, the use of these clusters is mainly for the task of channel assignment.

3.5.5 Location-Aided Routing (LAR)

The LAR Protocols use location information (which may be out of date by the time it is used) to reduce the search space for a desired route. Limiting the search space results in fewer route discovery messages.

3.5.5.1 Route Discovery Using Flooding

The possibility of using location information to improve performance of routing protocols for a MANET is discussed. As an illustration, we show how a route discovery protocol based on flooding can be improved. The route discovery algorithm using flooding is described next. When a node S needs to find a route to node D, node S broadcasts a route request message to all its neighbors; hereafter, node S will be referred to as the "sender," and node D as the "destination." A node (say, X), on receiving a route request message, compares the desired destination with its own identifier. If there is a match, it means that the request is for a route to itself (i.e., node X). Otherwise, node X broadcasts the request to its neighbors to avoid redundant transmissions of route requests, and node X only broadcasts a particular route request once (repeated reception of a route request is detected using sequence numbers). Figure 3.5 illustrates this algorithm. In Figure 3.5, node S needs to determine a route to node D. Therefore, node S broadcasts a route request to its neighbors.

When nodes B and C receive the route request, they forward it to all their neighbors. When node F receives the route request from B, it forwards the request to its neighbors. However, when node F receives the same route request from C, node F simply discards the route request. As the route request is propagated to various nodes, the path followed by the request is included in the route request packet. Using the above flooding algorithm, provided that the intended destination is reachable from the sender, the destination should eventually receive a route request message. On receiving the route request, the destination responds by sending a route reply message to the sender—the route reply message follows a path that is obtained by reversing the path followed by the route request received by D (the route request message includes the path traversed by the request).

It is possible that the destination will not receive a route request message (for instance, when it is unreachable from the sender, or route requests are lost due to transmission errors). In such cases, the sender needs to be able to reinitiate route discovery. Therefore, when a sender initiates route discovery, it sets a timeout. If during the timeout interval a route reply is not received, then a *new* route discovery

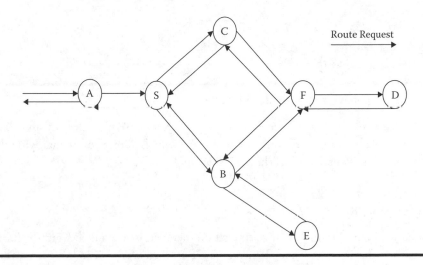

Figure 3.5 Illustration of Location Aided Routing (LAR).

is initiated (the route request messages for this route discovery will use a different sequence number than the previous route discovery—recall that sequence numbers are useful to detect multiple receptions of the same route request). Timeout may occur if the destination does not receive a route request, or if the route reply message from the destination is lost.

Route discovery is initiated either when the sender S detects that a previously determined route to node D is broken, or if S does not know a route to the destination. In our implementation, we assume that node S can know that the route is broken only if it attempts to use the route. When node S sends a data packet along a particular route, a node along that path returns a route error message, if the next hop on the route is broken. When node S receives the route error message, it initiates route discovery for destination D.

When using the above algorithm, observe that the route request would reach every node that is reachable from node S (potentially, all nodes in the ad hoc network). Using location information, we attempt to reduce the number of nodes to whom route request is propagated.

3.5.6 Ant Colony-Based Routing Algorithm (ARA)

ARA is a new on-demand routing algorithm for mobile, multihop ad hoc networks. The protocol is based on swarm intelligence and especially on the ant-colony-based meta-heuristic. These approaches try to map the solution capability of swarms to mathematical and engineering problems. This routing protocol is highly adaptive, efficient, and scalable. The main goal in the design of the protocol was to reduce the overhead for routing.

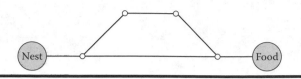

Figure 3.6   Behavior of individual ants.

The ant colony optimization meta-heuristic is a particular class of ant algorithms. Ant algorithms are multiagent systems which consist of agents with the behavior of individual ants, as shown in Figure 3.6.

3.5.6.1   Basic Ant Algorithm

The basic idea of the ant colony optimization metaheuristic is taken from the food-searching behavior of real ants. When ants are on their way to search for food, they start from their nest and walk toward the food. When an ant reaches an intersection, it has to decide which branch to take next. While walking, ants deposit pheromone, which marks the route taken. The concentration of pheromone on a certain path is an indication of its usage. With time, the concentration of pheromone decreases due to diffusion effects.

This property is important because it is integrating a dynamic method into the path-searching process. Figure 3.6 shows a scenario with two routes from the nest to the food place. At the intersection, the first ants randomly select the next branch. Because the lower route is shorter than the upper one, the ants which take this path will reach the food place first. On their way back to the nest, the ants again have to select a path. After a short time the pheromone concentration on the shorter path will be higher than on the longer path, because the ants using the shorter path will increase the pheromone concentration faster. The shortest path will thus be identified, and eventually all ants will only use this one.

This behavior of the ants can be used to find the shortest path in networks. Especially, the dynamic component of this method allows a high adaptation to changes in mobile ad hoc network topology, because in these networks the existence of links is not guaranteed and link changes occur very often.

The simple ant colony optimization meta-heuristic shown here illustrates different reasons why this kind of algorithm could perform well in mobile multihop ad hoc networks. We will discuss various reasons by considering important properties of mobile ad hoc networks.

> *Dynamic topology*: This property is responsible for the bad performance of several routing algorithms in mobile multihop ad hoc networks. The ant colony optimization meta-heuristic is based on agent systems and works with individual ants. This allows a high adaptation to the current topology of the network.

Local work: In contrast to other routing approaches, the ant colony optimization meta-heuristic is based only on local information, i.e., no routing tables or other information blocks have to be transmitted to neighbors or to all nodes of the network.

Link quality: It is possible to integrate the connection–link quality into the computation of the pheromone concentration, especially into the evaporation process. This will improve the decision process with respect to the link quality. It is here important to notice that the approach has to be modified so that nodes can also manipulate the pheromone concentration independent of the ants (i.e., data packets); for this, a node has to monitor the link quality.

Support for multipath: Each node has a routing table with entries for all its neighbors, which contains also the pheromone concentration. The decision rule, to select the next node, is based on the pheromone concentration on the current node, which is provided for each possible link. Thus, the approach supports multipath routing.

3.6 Hybrid Routing Protocols

Hybrid routing protocols are a new generation of protocol, which are both proactive and reactive in nature. These protocols are designed to increase scalability by allowing nodes with close proximity to work together to form some sort of a backbone to reduce the route discovery overheads. This is mostly achieved by proactively maintaining routes to nearby nodes and determining routes to faraway nodes using a route discovery strategy. Most hybrid protocols proposed to date are zone based, which means that the network is partitioned or seen as a number of zones by each node. Others group nodes are formed into trees or clusters. This section describes a number of different hybrid routing protocols proposed for MANETs.

- ■ ZRP
- ■ ZHLS
- ■ Scalable Location Updates Routing Protocol (SLURP)
- ■ Distributed Spanning Trees Based Routing Protocol (DST)
- ■ Distributed Dynamic Routing (DDR) Protocol

3.6.1 Zone Routing Protocol (ZRP)

The ZRP Protocol combines the advantages of the proactive and reactive approaches by maintaining an up-to-date topological map of a zone centered on each node. Within the zone, routes are immediately available. For destinations outside the zone, ZRP employs a route discovery procedure, which can benefit from the local routing information of the zones.

3.6.1.1 Motivation

Proactive routing uses excess bandwidth to maintain routing information, while reactive routing involves long route request delays. Reactive routing also inefficiently floods the entire network for route determination. ZRP aims to address the problems by combining the best properties of both approaches. ZRP can be classed as a hybrid reactive and proactive routing protocol.

In an ad hoc network, it can be assumed that the largest part of the traffic is directed to nearby nodes. Therefore, ZRP reduces the proactive scope to a zone centered on each node. In a limited zone, the maintenance of routing information is easier. Further, the amount of routing information that is never used is minimized. Still, nodes farther away can be reached with reactive routing. Because all nodes proactively store local routing information, route requests can be more efficiently performed without querying all the network nodes. Despite the use of zones, ZRP has a flat view over the network. In this way, the organizational overhead related to hierarchical protocols can be avoided. Hierarchical routing protocols depend on the strategic assignment of gateways or landmarks, so that every node can access all levels, especially the top level. Nodes belonging to different subnets must send their communication to a subnet that is common to both nodes. This may congest parts of the network. ZRP can be categorized as a flat protocol because the zones overlap. Hence, optimal routes can be detected and network congestion can be reduced. The behavior of ZRP is adaptive and depends on the current configuration of the network and the behavior of the users.

3.6.1.2 Architecture

The Zone Routing Protocol, as its name implies, is based on the concept of zones. A routing zone is defined for each node separately, and the zones of neighboring nodes overlap. The routing zone has a radius ρ _expressed in h hops. The zone thus includes the nodes, whose distance from the node in question is at most h hops. An example of a routing zone is shown in Figure 3.7, where the routing zone of S includes the nodes A–I, but not K. In the illustrations, the radius is marked as a circle around the node in question. It should, however, be noted that the zone is defined in hops, not as a physical distance.

The nodes of a zone are divided into peripheral nodes and interior nodes. Peripheral nodes are nodes whose minimum distance to the central node is exactly equal to the zone radius ρ. The nodes whose minimum distance is less than ρ are interior nodes. In Figure 3.7, the nodes A–F are interior nodes, the nodes G–J are peripheral nodes, and the node K is outside the routing zone. Note that node H can be reached by two paths, one with a length of two hops and one with a length of three hops. The node is, however, within the zone, because the shortest path is less than or equal to the zone radius. The number of nodes in the routing zone can be regulated by adjusting the transmission power of the nodes. Lowering the power

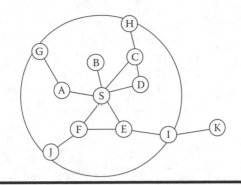

Figure 3.7   Example of a routing zone with ρ = 2.

reduces the number of nodes within direct reach and vice versa. The number of neighboring nodes should be sufficient to provide adequate reachability and redundancy. On the other hand, a too large coverage results in many zone members, and the update traffic becomes excessive. Further, large transmission coverage adds to the probability of local contention.

ZRP refers to the locally proactive routing component as the Intrazone Routing Protocol (IARP). The globally reactive routing component is named the Interzone Routing Protocol (IERP). IERP and IARP are not specific routing protocols. Instead, IARP is a family of limited-depth, proactive, link state routing protocols. IARP maintains routing information for nodes that are within the routing zone of the node. Correspondingly, IERP is a family of reactive routing protocols that offer enhanced route discovery and route maintenance services based on local connectivity monitored by IARP.

The fact that the topology of the local zone of each node is known can be used to reduce traffic when global route discovery is needed. Instead of broadcasting packets, ZRP uses a concept called "bordercasting." Bordercasting utilizes the topology information provided by IARP to direct query requests to the border of the zone. The bordercast packet delivery service is provided by the Bordercast Resolution Protocol (BRP). BRP uses a map of an extended routing zone to construct bordercast trees for the query packets. Alternatively, it uses source routing based on the normal routing zone. By employing query control mechanisms, route requests can be directed away from areas of the network that already have been covered.

To detect new neighbor nodes and link failures, ZRP relies on a Neighbor Discovery Protocol (NDP) provided by the MAC layer. NDP transmits Hello beacons at regular intervals. Upon receiving a beacon, the neighbor table is updated. Neighbors for which no beacon has been received within a specified time are removed from the table. If the MAC layer does not include an NDP, the functionality must be provided by IARP. The relationship between the components is illustrated in Figure 3.8. Route updates are triggered by NDP, which notifies IARP when the neighbor table is updated. IERP uses the routing table of IARP to

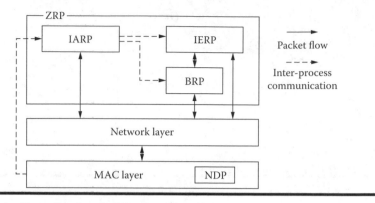

Figure 3.8 Zone Routing Protocol architecture.

respond to route queries. IERP forwards queries with BRP. BRP uses the routing table of IARP to guide route queries away from the query source.

3.6.1.3 Routing

A node that has a packet to send first checks whether the destination is within its local zone using information provided by IARP. In that case, the packet can be routed proactively. Reactive routing is used if the destination is outside the zone.

The reactive routing process is divided into two phases: the route request phase and the route reply phase. In the route request phase, the source sends a route request packet to its peripheral nodes using BRP. If the receiver of a route request packet knows the destination, it responds by sending a route reply back to the source. Otherwise, it continues the process by bordercasting the packet. In this way, the route request spreads throughout the network. If a node receives several copies of the same route request, these are considered as redundant and are discarded. The reply is sent by any node that can provide a route to the destination. To be able to send the reply back to the source node, routing information must be accumulated when the request is sent through the network. The information is recorded either in the route request packet or as next-hop addresses in the nodes along the path. In the first case, the nodes forwarding a route request packet append their address and relevant node or link metrics to the packet. When the packet reaches the destination, the sequence of addresses is reversed and copied to the route reply packet. The sequence is used to forward the reply back to the source. In the second case, the forwarding nodes record routing information as next-hop addresses, which are used when the reply is sent to the source. This approach can save transmission resources, as the request and reply packets are smaller. The source can receive the complete source route to the destination. Alternatively, the nodes along the path to the destination record the next-hop address in their routing table. In the bordercasting process, the bordercasting node sends a route request packet to each of its peripheral nodes. This type of one-to-many transmission can be implemented as a multicast to reduce resource usage. One approach is to let the source compute the multicast tree and attach routing instructions to the packet. This is called Root-Directed Bordercasting (RDB). Another approach is to reconstruct the tree at each node, whereas the routing instructions can be omitted. This requires that every interior node knows the topology seen by the bordercasting node. Thus, the nodes must maintain an extended routing zone with a radius of 2 (r − 1) hops. Note that in this case, the peripheral nodes where the request is sent are still at the distance r. This approach is named Distributed Bordercasting (DB). The zone radius is an important property for the performance of ZRP. If a zone radius of one hop is used, routing is purely reactive and bordercasting degenerates into flood searching. If the radius approaches infinity, routing is reactive. The selection of radius is a trade-off between the routing efficiency of proactive routing and the increasing traffic for maintaining the view of the zone.

3.6.1.4 Route Maintenance

Route maintenance is especially important in ad hoc networks, where links are broken and established as nodes move relatively to each other with limited radio coverage. In purely reactive routing protocols, routes containing broken links fail, and a new route discovery or route repair must be performed. Until the new route is available, packets are dropped or delayed.

In ZRP, the knowledge of the local topology can be used for route maintenance. Link failures and suboptimal route segments within one zone can be bypassed. Incoming packets can be directed around the broken link through an active multihop path. Similarly, the topology can be used to shorten routes, for example, when two nodes have moved within each other's radio coverage. For source-routed packets, a relaying node can determine the closest route to the destination that is also a neighbor. Sometimes, a multihop segment can be replaced by a single hop. If next-hop forwarding is used, the nodes can make locally optimal decisions by selecting a shorter path.

3.6.1.5 Query-Control Mechanisms

Bordercasting can be more efficient than flooding, because route request packets are only sent to the peripheral nodes, and thus only on the corresponding links. Further efficiency can be gained by utilizing multicast techniques. In that case, only one packet is sent on a link, although several peripheral nodes can reside behind this link. However, because the routing zones of neighboring nodes overlap, each node may forward route requests several times, which results in more traffic than in flooding. When a node bordercasts a query, the complete routing zone is effectively covered. Any further query messages entering the zone are redundant

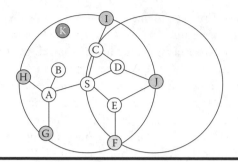

Figure 3.9   Query detection example.

and result in wasted transmission capacity. The excess traffic is a result from queries returning to covered zones instead of covered nodes as in traditional flooding. To solve this problem, ZRP needs query-control mechanisms, which can direct queries away from covered zones and terminate query packets before they are delivered to peripheral nodes in regions of the network already covered by the query.

ZRP uses three types of query-control mechanisms: query detection, early termination, and random query-processing delay. Query detection caches the queries relayed by the nodes. With early termination, this information is used to prune bordercasting to nodes already covered by the query.

3.6.1.6   Query Detection

When a bordercast is issued, only the bordercasting node is aware that the routing zone is covered by the query. When the peripheral nodes continue the query process by bordercasting to their peripheral nodes, the query may be relayed through the same nodes again. To illustrate with an example, the node S in Figure 3.9 bordercasts a query to its peripheral nodes F–J. As the node J continues by bordercasting to the nodes C, S, and E, the query is again relayed by nodes D and E. The query issued by node J to nodes C, S, and E is redundant, because these nodes have been covered by the previous query.

To be able to prevent queries from reappearing in covered regions, the nodes must detect local query-relaying activity. BRP provides two query detection methods: QD1 and QD2. Firstly, the nodes that relay the query are able to detect the query (QD1). Secondly, in single-channel networks, it is possible to listen to the traffic by other nodes within the radio coverage (QD2). Hence, it is possible to detect queries relayed by other nodes in the zone. QD2 can be implemented by using IP broadcasts to send route queries. Alternatively, a unicast can be used if the MAC and IP layers operate in promiscuous mode.

In the above example, all nodes except node B relay the query of S. They are thus able to use QD1. Node B does not belong to the bordercast tree, but it is able to overhear the relayed query using QD2. However, node K does not overhear the

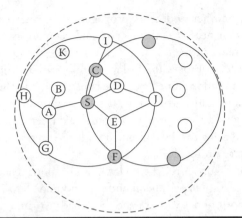

Figure 3.10   The extended routing zone of E.

message, and is therefore unaware that the zone of node S is covered. A query detection table is used to cache the detected queries. For each entry, the cache contains the address of the source node and the query ID. The address–ID pair is sufficient to uniquely identify all queries in the network. The cache may also contain other information depending on the query detection scheme. In particular, the address of the node that most recently bordercasted a query is important.

3.6.1.7   Early Termination

With early termination (ET), a node can prevent a route request from entering already covered regions. Early termination combines information obtained through query detection with the knowledge of the local topology to prune branches leading to peripheral nodes inside covered regions. These regions consist of the interior nodes of nodes that already have bordercasted the query. A node can also prune a peripheral node if it has already relayed a query to that node. Early termination requires topology information extending outside the routing zone of the node. The information is required to reconstruct the bordercast tree of other nodes within the routing zone. The extended routing zone has a radius of 2 (r − 1). Alternatively, in the case of root-directed bordercasting (RDB), the topology of the standard routing zone and information about cached bordercast trees can be used. In the previous example, node E can use the information in its query detection table to prune the query that the node J sends to its peripheral node F. Node E has an extended routing zone with radius 2 (r − 1) = 3, shown as a dashed circle in Figure 3.10.

3.6.1.8   Random Query-Processing Delay

When a node issues a node request, it takes some time for the query to be relayed along the bordercast tree and to be detected through the query detection mechanisms.

During this time, another node may propagate the same request. This can be a problem when several nearby nodes receive and rebroadcast a request at roughly the same time. To reduce the probability of receiving the same request from several nodes, a Random Query-Processing Delay (RQPD) can be employed. Each border-casting node waits a random time before the construction of the bordercast tree and the early termination. During this time, the waiting node can detect queries from other bordercasting nodes and prune the bordercast tree. To avoid additional route discovery delay, the delay can be combined with the pretransmission jitter used by many route discovery protocols.

Assume that in Figure 3.10, the nodes C and S both receive a query. Node C schedules a bordercast to its peripheral node E, and node S to its peripheral node F. Without RQPD, both nodes would issue the broadcast simultaneously, and there-after detect the message of the neighbor node. With RQPD, the node C may detect the query sent by node S during the delay, and prune the branch leading to E.

3.6.1.9   Caching

Caching is a technique used for reducing control traffic. The nodes cache active routes, and by using this cache, the frequency of route discovery procedures can be reduced. Changes in network topology, such as broken links, are compensated by local path repair procedures. A new path then substitutes the path between the ends of the broken link, and a path update message is sent to the endpoints of the path. Because the repair reduces the efficiency of the routes, the endpoints may initiate a new route discovery procedure after a number of repairs.

3.6.2   Zone-Based Hierarchical Link State (ZHLS)

A "peer-to-peer" hierarchical routing protocol, Zone-Based Hierarchical LSR Protocol (ZHLS), incorporates location information into a novel peer-to-peer hierarchical routing approach. The network is divided into no overlapping zones. Aggregating nodes into zones conceals the detail of the network topology. Initially, each node knows its own position and therefore zone ID through GPS. After the network is established, each node knows the low-level (node-level) topology about node connectivity within its zone and the high-level (zone-level) topology about zone connectivity of the whole network. A packet is forwarded by specifying the hierar-chical address—zone ID and node ID—of a destination node in the packet header. Unlike other hierarchical protocols, there are no clusterheads in this protocol. The high-level topological information is distributed to all nodes (i.e., in a peer-to-peer manner). This peer-to-peer characteristic avoids traffic bottleneck, prevents a single point of failure, and simplifies mobility management. Similar to ZRP, ZHLS is a hybrid reactive and proactive scheme. It is proactive if the destination is within the same zone of the source. Otherwise, it is reactive because a location search is needed to find the zone ID of the destination. However, unlike ZRP, ZHLS requires GPS and maintains a high-level hierarchy for interzone routing. Location search is performed by unicasting one location request to each zone. Routing is done by specifying the zone ID and the node ID of the destination, instead of specifying an ordered list of all the intermediate nodes between the source and the destination. Intermediate link breakage may not cause any subsequent location search. Because the network consists of nonoverlapping zones in ZHLS, frequency reuse is readily deployable in ZHLS.

3.6.2.1 Zone Map

The network is divided into zones under ZHLS. A node knows its physical location by geolocation techniques such as GPS; then, it can determine its zone ID by mapping its physical location to a zone map, which has to be worked out at the design stage. The zone size depends on factors such as node mobility, network density, transmission power, and propagation characteristics. The partitioning can be based on simple geographic partitioning or on radio propagation partitioning. The geographic partitioning is much simpler and does not require any measurement of radio propagation characteristics, whereas the radio propagation partitioning is more accurate for frequency reuse. Radio propagation partitioning is preferable if a propagation measurement can be done at the design stage. However, some applications, such as emergency disaster rescue operations, tactical military communications, and law enforcement, do not permit such measurements. In such cases, a simple geographic partitioning has to be used.

3.6.2.2 Hierarchical Structure of ZHLS

Two levels of topology are defined in ZHLS: node-level topology and zone-level topology. If any two nodes are within the communication range, a physical link exists. The node-level topology (Figure 3.11) provides the information on how the nodes are connected by these physical links.

For example, in Figure 3.11, if node a wants to send a data packet to node i, the data has to pass through a–b–c–f. If there is at least one physical link connecting any two zones, a virtual link then exists. The zone-level topology (Figure 3.12) tells how the zones are connected by these virtual links. For example, in Figure 3.12, the virtual links between zone 4 and zone 3 are 4–1–3.

We will see how a node uses the node-level topology to route a packet within a zone and how it uses the zone-level topology to route a packet between the zones. To facilitate this hierarchical LSR protocol, each node receives two types of link state packets (LSPs): node LSPs and zone LSPs. The node LSP of a particular node contains a list of its connected neighbors and is propagated locally within its zone.

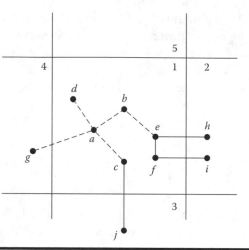

Figure 3.11 Node-level topology.

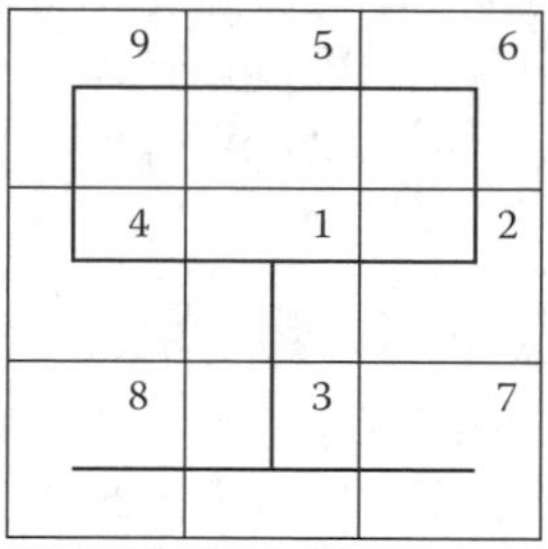

9	5	6	
4	1	2	
8	3	7	

Figure 3.12 Zone-level topology.

The zone LSP contains a list of its connected zones and is propagated globally throughout the network.

3.7 Summary

Mobile networks can be classified into infrastructure networks and mobile ad hoc networks according to their dependence on fixed infrastructures. In an infrastructure mobile network, mobile nodes have wired access points (or base stations) within their transmission range. The access points compose the backbone for an infrastructure network. In contrast, mobile ad hoc networks are autonomously self-organized networks without infrastructure support. The routing architecture of self-organized networks can be either hierarchical or flat. In most self-organized networks, the hosts will be acting as independent routers, which implies that routing architecture should conceptually be flat, that is, each address serves only as an identifier and does not convey any information about one host being topologically located with respect to any other node. In a flat self-organized network, the mobility management is not necessary because all of the nodes are visible to each other via routing protocols.

All existing routing protocols assume that all mobile hosts have the same properties based on the spirit of a self-organized network as a collection of "equal" peers opportunistically using each other's services to communicate. Although this is true in some circumstances, there are such situations where the network will include hosts with preponderant bandwidth, guaranteed power supply, and high-speed wireless links. Such hosts are referred to as SuperHosts.

QoS routing provides support to effectively control the total traffic that can flow into the network. QoS routing is a routing mechanism under which paths for flows are determined according to resource availability in the network as well as the QoS requirement of flows. QoS routing means that it selects routes with sufficient resources for the requested QoS parameters. Multicast routing sends a single copy of a data packet simultaneously to multiple receivers over a communication link that is shared by the paths to the receivers. The sharing of links in the collection of the paths to receivers implicitly defines a tree used to distribute multicast packets. In contrast to unicast routing, multicast routing is a very useful and efficient way to support group communication. This is especially the case in self-organized networks where bandwidth is limited and energy is constrained. In addition, a self-organized network often consists of several cooperative work groups.

A proactive routing protocol is also called a "table-driven" routing protocol. Using a proactive routing protocol, nodes in a mobile ad hoc network continuously evaluate routes to all reachable nodes and attempt to maintain consistent, up-to-date routing information. Therefore, a source node can get a routing path immediately if it needs one.

Reactive routing protocols for mobile ad hoc networks are also called "on-demand" routing protocols. In a reactive routing protocol, routing paths are searched only when needed. A route discovery operation invokes a route determination procedure. The discovery procedure terminates when either a route has been found or no route is available after examination for all route permutations. Hybrid routing protocols are proposed to combine the merits of both proactive and reactive routing protocols and overcome their shortcomings. Normally, hybrid routing protocols for mobile ad hoc networks exploit hierarchical network architectures. Proper proactive routing approaches and reactive routing approaches are exploited in different hierarchical levels, respectively.

3.8 Problems

3.1 Explain the classification of mobile ad hoc networks based on infrastructure.

3.2 Discuss briefly the design issues in developing a routing protocol.

3.3 List the factors that will make wireless links unidirectional in ad hoc networks.

3.4 Give the suitable reasons why a traditional multicast routing protocol cannot be used in ad hoc networks.

3.5 Explain a proactive routing protocol with suitable examples.

3.6 Describe a Wireless Routing Protocol with a ncat diagram.

3.7 Discuss how routing is done using the Destination-Sequenced Distance Vector Protocol.

3.8 Explain how multipoint relays are used in the Optimized Link State Routing Protocol.

3.9 How is message reduction done in Fisheye State Routing?

3.10 Explain a reactive routing protocol with an example.

3.11 Explain the Ad Hoc On-Demand Distance Vector Routing Protocol with a proper illustration.

3.12 Describe dynamic source routing used in ad hoc networks.

3.13 How is the Temporally Ordered Routing Algorithm used to route the packets in ad hoc networks?

3.14 Explain how the Cluster-Based Routing Protocol reduces overhead packets in ad hoc networks.

3.15 Discuss Location-Aided Routing with an appropriate example.

3.16 With a suitable real-time example, explain the Ant-Colony-Based Routing Algorithm.

3.17 Explain hybrid routing protocols with an example.

3.18 Explain the Zone Routing Protocol with a neat diagram.

3.19 Discuss the Zone-Based Hierarchical Link State Routing Protocol deployed in ad hoc networks.

3.20 Describe the Scalable Location Updates Routing Protocol using a suitable example.

3.21 How does a Distributed Spanning Trees Based Routing Protocol work well in ad hoc networks, and what are its demerits?

3.22 Explain distributed dynamic routing with a suitable example.

Bibliography

M. Abolhasan, T. Wysocki, and E. Dutkiewicz, A review of routing protocols for mobile ad hoc networks, *Elsevier Journal of Ad Hoc Networks*, 1–22, 2004.

A. Adams, J. Nicholas, and W. Siadak, Protocol Independent Multicast—Dense Mode (PIM-DM): Protocol specification (revised), IETF draft, February 2003.

S. Basagni, I. Chlamtac, V. R. Syrotivk, and B. A. Woodward, A distance effect algorithm for mobility (DREAM), in Proceedings of the Fourth Annual ACM/IEEE International Conference on Mobile Computing and Networking (Mobicom'98), Dallas, TX, 1998.

B. Bellur, R. G. Ogier, and F. L Templin, Topology broadcast based on reverse-path for-warding routing protocol (TBRPF), Internet draft, draft-ietf-manet-tbrpf-06.txt, work in progress, 2003.

S. Corson and J. Macker, RFC 2501: Mobile ad hoc networking (MANET): Rout-ing protocol performance issues and evaluation considerations, Internet draft, draft-ietf-manet-issues-01.txt.

M. S. Corson and A. Ephremides, A distributed routing algorithm for mobile wireless networks, ACM/Baltzer *Wireless Networks*, 1(1):61–81, 1995.

S. Das, C. Perkins, and E. Royer, Ad hoc on demand distance vector (AODV) routing, Internet draft, draft-ietf-manetaodv-11.txt, work in progress, 2002.

S. Das, C. Perkins, and E. Royer, Performance comparison of two on-demand routing protocols for ad hoc networks, INFOCOM 2000, pp. 3–12.

B. Fenner, M. Handley, H. Holbrook, and I. Kouvelas, Protocol Independent Multicast—Sparse Mode (PIM-SM): Protocol specification (revised), IETF draft, December 2002.

L. R. Ford and D. R. Fulkerson, *Flows in networks*, Princeton University Press, Princeton, NJ, 1962.

M. Gerla, Fisheye state routing protocol (FSR) for ad hoc networks, Internet draft, draft-ietf-manet-aodv-03.txt, work in progress, 2002.

M. Geunes, U. Sorges, and I. Bouazizi, ARA: The ant-colony based routing algorithm for MANETs, in ICPP Workshop on Ad Hoc Networks (IWAHN 2002), August 2002, pp. 79–85.

Z. J. Hass, R. Pearlman, Zone routing protocol for ad-hoc networks, Internet draft, draft-ietf-manet-zrp-02.txt, work in progress, 1999.

C. Hedrick, RFC 1058: Routing information protocol, Network Working Group, June 1988.

IETF Manet charter, http://www.ietf.org/html.charters/manet-charter.html.

P. Jaquet, P. Muhlethaler, T. Clausen, A. Laouiti, A. Qayyum, and L. Viennot, Optimized link state routing protocol for ad hoc networks, IEEE INMIC, Pakistan, 2001.

M. Jiang, J. Ji, and Y. C. Tay, Cluster based routing protocol, Internet draft, draft-ietf-manet-cbrp-spec-01.txt, work in progress, 1999.

M. Joa-Ng and I-T. Lu, A peer-to-peer zone-based two-level link state routing for mobile ad hoc networks, *IEEE Journal on Selected Areas in Communications*, 17(8):1415–1425, 1999.

D. Johnson, D. Maltz, and J. Jetcheva, The dynamic source routing protocol for mobile ad hoc networks, Internet draft, draft-ietf-manet-dsr-07.txt, work in progress, 2002.

Y. B. Ko and N. H. Vaidya, Location Aid Routing (LAR) in mobile ad hoc networks, in Proceedings of the ACM/IEEE MOBICOM, Oct. 1998.

C. Liu and J. Kaiser, A survey of mobile ad hoc network routing protocols, University of Ulm Technical Report Series, No. 2003-08, University of Ulm, Germany, 2005.

Y. Lu, W. Wang, Y. Zhong, and B. Bhargava, Study of distance vector routing protocols for mobile ad hoc networks, in Proceedings of the First IEEE International Conference on Pervasive Computing and Communications (PerCom'03), 2003.

R. A. Meyer, PARSEC User Manual, UCLA Parallel Computing Laboratory, Los Angeles, CA, http://pcl.cs.ucla.edu.

J. Moy, RFC 2328: OSPF Version 2, Network Working Group, April 1998.

J. Moy, RFC 1584: Multicast open shortest path first (MOSPF), Network Working Group, March 1994.

S. Murthy and J. J. Garcia-Luna-Aceves, A routing protocol for packet radio networks, in Proceedings of the First Annual ACM International Conference on Mobile Computing and Networking, Berkeley, CA, 1995, pp. 86–95.

N. Nikaein, C. Bonnet, and N. Nikaein, HARP: Hybrid Ad Hoc Routing Protocol, in Proceedings of IST 2001: International Symposium on Telecommunications, Tehran, Iran, 2001.

N. Nikaein, H. Laboid, and C. Bonnet, Distributed dynamic routing algorithm (DDR) for mobile ad hoc networks, in Proceedings of the MobiHOC 2000: First Annual Workshop on Mobile Ad Hoc Networking and Computing, 2000.

NS-2, http://www.isi.edu/nsnam/ns/.

OPNET Inc., http://www.opnet.com.

V. Park and S. Corson, Temporally-Ordered Routing Algorithm (TORA) version 1 functional specification, ETF Internet draft, 1997.

G. Pei, M. Gerla, X. Hong, and C. Chiang, A wireless hierarchical routing protocol with group mobility, in Proceedings of Wireless Communications and Networking, New Orleans, LA, 1999.

C. E. Perkins and T. J. Watson, Highly dynamic destination sequenced distance vector routing (DSDV) for mobile computers, in ACM SIGCOMM '94 Conference on Communications Architectures, London, 1994.

QualNet simulator software, Scalable Networks Inc., http://www.scalable-networks.com.

S. Radhakrishnan, N. S. V. Rao, G. Racherla, C. N. Sekharan, and S. G. Batsell, DST: A routing protocol for ad hoc networks using distributed spanning trees, in IEEE Wireless Communications and Networking Conference, New Orleans, LA, 1999.

R. Ramanathan and M. Steenstrup, Hierarchically-organized, multihop mobile wireless networks for quality-of-service, ACM/Baltzer *Mobile Networks and Applications*, 3(1):101–119.

E. Royer and C-K. Toh, A review of current routing protocols for ad hoc mobile wireless networks, *IEEE Personal Communications*, 6:46–55, April 1999.

P. Sinha, R. Sivakumar, and V. Bharghaven, CEDAR: A core-extraction distributed ad hoc routing algorithm, IEEE INFOCOM, March 1999.

A. S. Tanenbaum, *Computer networks*, 3rd edition, Prentice Hall, Upper Saddle River, NJ, 1996.

D. Waitzman, C. Partridge, and S. Deering, RFC 1075: Distance vector multicast routing protocol, Network Working Group, November 1988.

S-C. Woo and S. Singh, Scalable routing protocol for ad hoc networks, *Wireless Networks*, 7(5):513–529, 2001.

Chapter 4

Multicast Routing Protocols for Mobile Ad Hoc Networks

4.1 Introduction

Mobile ad hoc networks (MANETs) have numerous practical applications, such as emergency and relief operations, military exercises and combat situations, and conference or classroom meetings. Each of these applications can potentially involve different scenarios, with movement pattern, density, and traffic rate dependent on the environment and the nature of the interactions among the participants. For example, in a search-and-rescue operation, individuals may fan out to search a wide area, resulting in a fairly regular pattern of movement, low density, and a low traffic rate. In a battlefield scenario, the movements of soldiers may be heavily influenced by the movements of their commander, with higher density and a higher traffic rate. In other cases, the environment itself may give rise to movement patterns and density such as patrons visiting an exhibit hall and moving among a selected group of displays. In addition, depending upon the communication need, applications can be very demanding, requiring the system to support very high traffic rates.

To enable group communication in these scenarios, a number of ad hoc network routing protocols have been proposed. In making the transition from wired to wireless networking, protocol designers have focused on the obvious challenges of designing a multicast routing protocol that can cope with a mobile environment. As a result, the main goal of most ad hoc multicast protocols is to build and maintain a multicast tree or mesh in the face of a mobile environment, with fast reactions to network changes so that the packet loss is minimized.

4.2 Issues in Designing a Multicast Routing Protocol

Many unique characteristics of MANETs have posed new challenges in multicast routing protocol design: dynamic network topology, energy constraints, lack of network scalability and a centralized entity, and the different characteristics between wireless links and wired links such as limited bandwidth and poor security.

There are several issues involved here, which are discussed below:

Robustness: Because nodes will be moving, link failures are common in MANETs. Data sent by a source may be dropped, which results in a low packet delivery ratio. Hence, a multicast routing protocol should be robust enough to withstand the mobility of nodes and achieve a high packet delivery ratio.

Efficiency: In an ad hoc network environment, where the bandwidth is scarce, the efficiency of the multicast routing protocol is very important. "Multicast efficiency" is defined as the ratio of the total number of data packets received by the receivers to the total number of data and control packets transmitted in the network.

Control overhead: To keep track of the members in a multicast group, the exchange of control packets is required. This consumes a considerable amount of bandwidth. Because bandwidth is limited in ad hoc networks, the design of a multicast protocol should ensure that the total number of control packets transmitted for maintaining the multicast group is kept to a minimum.

Quality of service: MANETs, typically scarce and dynamic network topology, make it difficult to perform efficient resource utilization or to execute critical real-time applications in such environments. Based on this consideration, it is necessary to provide Quality-of-Service (QoS) multicast routing support to effectively control the total traffic that can flow into the network. QoS multicast routing is a routing mechanism under which paths for flows are determined according to resource availability in the network as well as the QoS requirement of flows. QoS multicast routing means that it selects routes with sufficient resources for the requested QoS parameters. The goal of QoS multicast routing has two points. The first one is to meet the QoS requirements for each admitted connection, and the second one is to achieve global efficiency in resource utilization. Thus, QoS routing will consider multiple constraints and provide better load balance by allocating traffic on different paths, subject to the QoS requirement of different traffic.

Dependency on the unicast routing protocol: If a multicast routing protocol needs the support of a particular routing protocol, then it is difficult for the multicast protocol to work in heterogeneous networks. Hence, it is desirable if the multicast routing protocol is independent of any specific unicast routing protocol.

Resource management: Ad hoc networks consist of a group of mobile nodes, with each node having limited battery power and memory. An ad hoc multicast routing protocol should use minimum power by reducing the number of packet transmissions. To reduce memory usage, it should use minimum state information.

4.3 Classification of Multicast Routing Protocols

Multicast routing protocols for MANETs can be broadly classified into two types: application-independent and application-dependent multicast protocols. Although application-dependent multicast protocols are used for conventional multicasting, they are meant only for specific applications for which they are designed. Application-independent multicast protocols can be classified as follows.

4.3.1 Based on Topology

Based on multicast topology, ad hoc multicast routing can be classified into two types: tree based and mesh based.

4.3.1.1 Tree-Based Multicast

Tree-based multicast is generally used in wired and infrastructure mobile networks (i.e., mobile networks with base stations) as well as in MANETs. Figure 4.1 (a) and (b) shows an example of a multicast tree. The tree consists of a root node (r), three intermediate nodes (p, s, and t), seven member nodes of a multicast group, and ten tree links. A multicast packet is delivered from the root node r to seven group members. For node u, for instance, the packet transmission is relayed through two tree links, that is, from r to q and then q to u. This requires two transmissions and two receives. Now consider the last transmission from q to u. Even though all nodes within node q's radio transmission range can receive the multicast packet, only node u will receive the packet because the rest of the nodes are not addressed.

To maintain the tree structure even when nodes move, group members periodically send Join Requests to the root node so that the multicast tree can be updated using the path information included in the Join Request messages. Joining a multicast group causes reports (i.e., join messages) to be periodically sent, whereas leaving a multicast group does not lead to any explicit action. The period must be carefully chosen to balance the overhead associated with the tree update and the delay caused

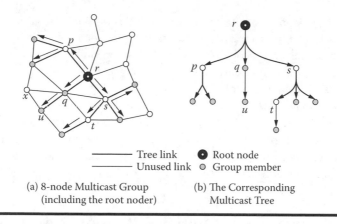

—————— Tree link	●	Root node
————— Unused link	○	Group member

(a) 8-node Multicast Group   (b) The Corresponding
(including the root noder)   Multicast Tree

Figure 4.1   Tree-based multicasting routing.

by the tree not being updated in a timely fashion when nodes move. Depending
on the number of trees per multicast group, a tree-based multicast can be further
classified as a per-source tree multicast and shared tree multicast. Although a per-
source tree multicast is established and maintained for each source node of a multicast group, shared tree multicast utilizes a single shared tree for all multicast source
nodes. In the per-source tree multicast, each multicast packet is forwarded along
the most efficient path from the source node to each and every multicast group
member, but this method incurs a lot of control overhead to maintain many trees.
On the other hand, the shared tree multicast has lower control overhead because
it maintains only a single tree for a multicast group and thus is more scalable.
However, the path is not necessarily optimal and the root node is easily overloaded
due to the sharing of a single tree.

4.3.1.2   Mesh-Based Multicast

Tree-based protocols, however, may not perform well in the presence of highly
mobile nodes because multicast tree structure is fragile and needs to be frequently
readjusted as the connectivity changes. A new approach unique to MANETs is the
mesh-based multicast.

    A mesh is different from a tree because each node in a mesh can have multiple
parents. Using a single mesh structure spanning all multicast group members, multiple links exist and other links are immediately available when the primary link is
broken due to node mobility. This avoids frequent network reconfigurations, which
minimizes disruptions to ongoing multicast sessions and reduces the control overhead
to reconstruct and maintain the network structure.

    Figure 4.2a,b shows an example of a mesh-based multicast. It includes six redun-
dant links in addition to ten tree links. A multicast packet is broadcast within a
multicast mesh. Thus, sending a packet from R to U involves three transmissions

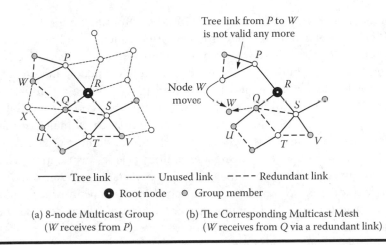

——— Tree link ┄┄┄ Unused link ─ ─ ─ Redundant link

● Root node ○ Group member

(a) 8-node Multicast Group (b) The Corresponding Multicast Mesh
  (*W* receives from *P*)   (*W* receives from *Q* via a redundant link)

Figure 4.2 Mesh-based multicasting routing.

(R, Q, U) and fourteen receives (5 neighbors of R, 6 neighbors of Q, and 3 neighbors of U). For example, the transmission from node Q is received not only by U but also by neighbor nodes R, S, T, W, and X; the redundant link from Q to W may be useful when the path from P to W is broken, as shown in Figure 4.2 (b).

4.3.2 Based on Initialization of the Multicast Session

The multicast group formation can be initiated by the source as well as by the receivers. In a multicast protocol, if the group formation is initiated only by the source node, then it is called a "source-initiated multicast routing protocol," and if it is initiated by the receivers of the multicast group, it is called a "receiver-initiated multicast routing protocol."

4.3.3 Based on Topology Maintenance Mechanism

Maintenance of the multicast topology can be done either by the soft-state approach or by the hard-state approach. In the soft-state approach, control packets are flooded periodically to refresh the route, which leads to a high packet delivery ratio at the cost of more control overhead, whereas in the hard-state approach, the control packets are transmitted only when a link breaks, resulting in lower control overhead, but at the cost of a low packet delivery ratio.

4.3.4 Based on Zone Routing

In Zone-Based Multicast Routing Protocol (ZBMRP), when a node has multicast packets to send but no route information is available, it starts to create a forwarding mesh in the entire network as On-Demand Multicast Routing Protocol (ODMRP) does. Then, it creates multiple mesh-based routing zones, including source and branch zones, along the route from the source node to multicast receiver nodes according to the distribution of the source node, receiver nodes, and forwarding group nodes in the forwarding mesh. Zone leaders are selected according to the First Declaration Wins (FDW) principle which is responsible for creating and maintaining zones periodically. Inside each zone, a mesh-based multicast routing strategy similar to ODMRP is used. Zone size and the number of zones can be decided according to the network size and multicast nodes distribution. Tunneling technology is employed to deliver multicast packets among zones and other sporadic multicast receivers that are not included in any zone in which multicast packets are encapsulated in the unicast packet for transmission. Because control packets flooding is restricted inside multicast zones, multicast overhead will be vastly reduced, and good scalability can be obtained.

The ZBMRP scheme integrates three advanced techniques—mesh based, on demand, and zone based, such as ODMRP, Ad Hoc On-Demand Distance Vector (AODV), and Zone Routing Protocol (ZRP), respectively—and has the characteristics of being adaptive to network topology change, being robust to nodal mobility, and having good scalability. ZBMRP will provide adequate multicast service to MANETs where bandwidth is limited, topology changes frequently, and power is constrained.

4.3.4.1 Protocol Overview: Mesh Establishment Phase

When a multicast source has packets to send but cannot find any route and group membership information, it broadcasts a member advertising and route request packet, termed "RREQ," to the entire network. Only multicast receivers send back a route reply message, called "RREP," to allow the source node to get the current routing information. Then ZBMRP uses the same strategy as ODMRP does to establish a mesh of nodes for forwarding packets between a multicast source and receivers.

The mesh is created using the forwarding group concept. The forwarding group is a set of nodes that are in charge of forwarding multicast packets. It supports the shortest paths between any member pairs. A multicast receiver may serve as a forwarding group node if it is on the path between a multicast source and another receiver.

The ZBMRP further classifies forwarding group nodes into two categories: FG-F and FG-B. FG-F means a forwarding group node which only forwards packets to one other node. FG-B is a forwarding group node that should forward packets to more than one node (i.e., forwarding branches), as seen with node C (shown in Figure 4.1). We call the source node "FG-B" and the multicast receivers "Zone-Associated Nodes" (ZANs). For example, in Figure 4.3, nodes B, C, R1, and R6 are

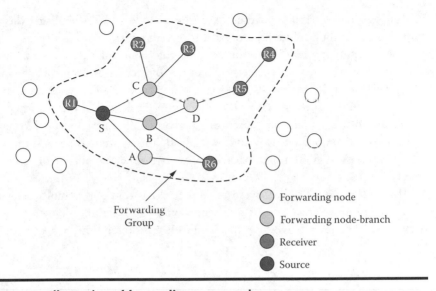

Figure 4.3 Illustration of forwarding group nodes.

Figure 4.3 Illustration of forwarding group nodes.

the nearest downstream ZANs of source node S. Node A is an FG-F node but not a ZAN.

4.3.4.2 *Source Zone Creation*

After the forwarding mesh between a multicast source and receivers is established in the entire network, mesh-based multicast routing zones are created according to the distribution of the source node, FG-B's, and multicast receiving member nodes, that is, ZANs.

A multicast source node will first establish a mesh-based multicast zone, named a source zone. It collects the information of its nearest downstream ZANs from RREP messages that it receives. Such information includes an Internet Protocol (IP) address and distance (in terms of hop counts). If the source's nearest downstream ZAN is far away (e.g., more than three hops away) and relatively sparse, then the source node will not establish a source zone (or we can say the zone size is 0). It just tunnels multicast packets in the unicast packets to its nearest downstream ZANs. If the source node finds many ZANs within N hops, then it establishes a source zone with a zone radius of N. The source node becomes the leader of this source zone. A zone leader is in charge of constructing and maintaining a zone. When N is no less than the size of the entire network, ZBMRP becomes ODMRP.

After the source node selects the size of the source zone as N, it sets the Time-to-Live (TTL) of the periodically flooded RREQ packet to be N, puts the zone ID (i.e., the IP address of the zone leader) in the reserved field of the IP header, and then sends out the IP multicast packet. Inside the source zone, forwarding mesh-based routing strategy is used. The source node and ZANs inside the zone communicate through a mesh of forwarding group nodes.

For the sporadic ZANs of a source node that is outside the source zone, the source node will tunnel packets to them. An FG-B receiving packet from the source node with zone information just joins the source zone as a normal zone member; it will not build a zone by itself. These receiving packets find out whether their ZANs are in the source zone based on the source zone size (optimal zone size depends largely on node density and traffic load, and its study is beyond the scope of this work). An FG-B tunnels packets to its ZANs that are outside of the source zones, if there are any.

If a downstream ZAN receives the same packets from both a multicast source node and an upstream FG-B, it will send a message N-Tunnel (Not to Tunnel) packet to its upstream FG-B to notify the FG-B to stop sending packets to it.

4.3.4.3 Branch Zone Creation

If an FG-B gets data packets without zone information, which means it gets the packets through tunneling, it is outside the upper-level zone and has to create and maintain its mesh-based routing zone according to the distribution of its nearest ZANs. We call this kind of zone a "branch zone." Any other FG-B inside this zone just joins it and will not create the zone of itself, as an FG-B source zone does. It is a kind of FDW strategy. This kind of work will continue until the far end of the network.

Two or more FG-B's do not contend for building a same branch zone. If a ZAN resides in multiple zones, it will receive multiple copies of the same multicast packet from several zone leaders with different zone IDs. It then just discards the replicate packet(s). Note that if this ZAN received a replicate packet tunneled from its upstream ZAN, it needs to send an N-Tunnel message packet to its upstream ZAN to save bandwidth. Upon receiving data packets without zone information, an FG-B will start to establish its zone. Thus, along with the data packets being delivered, multiple mesh-based routing zones will be established along the path from the source node to multicast receivers.

In Figure 4.4, we show an example of several zones being created. Node S is the source node. The source zone has a zone radius of 1. The branch zone created by node C has a zone radius of 2. Note that the dashed lines represent tunneling transmissions. Therefore, nodes R2 and R3 receive their multicast data through tunneling.

4.3.4.4 Zone and Route Maintenance

A zone leader periodically broadcasts RREQ messages inside the zone with the TTL equal to N (zone radius). Every node inside the zone that receives the RREQ forwards

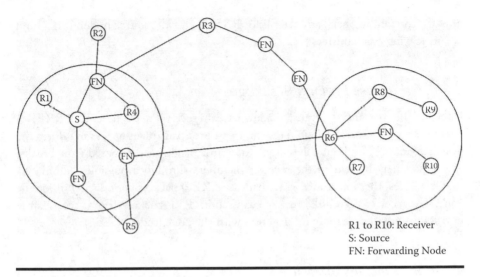

R1 to R10: Receiver
S: Source
FN: Forwarding Node

Figure 4.4   Zone creation.

the RREQ until the TTL becomes zero. It also sends the RREP back to the zone leader. With these, the zone leader updates the mesh routing inside the zone.

For the sporadic ZANs that are not included in any zone, the upstream ZAN of it tunnels packets to it unless the upstream ZAN receives explicit N-Tunnel messages from these sporadic ZANs. A periodic positive update of tunneling membership is also possible. In this technique, the downstream nodes periodically send the upstream nodes a tunnel message. Reception of such messages indicates validity of the tunnel.

4.3.4.5 New Node Joining the Multicast Group

During the process of multicasting, if a new node wants to join the multicast group, it explicitly generates an Route Repair with a Repair (RREP-R) flag packet that will be broadcasted to its neighbors. This RREP-R message will be forwarded until it is received by a forwarding group node, or source node. Then this node will be added to the multicast group, through a zone or a forwarding route.

If the RREP-R is terminated at the normal forwarding group node with no branch (i.e., FG-F), then this FG-F node will become as an FG-B node which will update this information to its upstream ZAN and be responsible for forwarding packets to this new group member.

4.3.4.6 Multicast Group Member Leaving the Group

Only the multicast group member node outside any zone needs to send an N-Tunnel message to its upstream ZAN to tell it to leave. Other group member nodes need only to stop sending back any RREPs or RREQs (for the source node). It is a kind of soft leaving mechanism.

4.3.4.7 Process for Link Breakage

If the link for forwarding breaks, the downstream receiver needs to send an RREP-R packet as a new node does to join the multicast group again, as well as to inform the downstream ZANs about this state. If one link inside a mesh-based zone breaks, the packet may be sent to the receiver members through a possible redundancy route. Besides, the zone leader will broadcast RREQ packets periodically inside the zone. The receiver with link breakage can send RREP packets after it receives the new RREQ to update route information with the zone leader.

4.3.4.8 Unicast Capability

A unicast can be seen as a special case of ZBMRP (i.e., there is only one group receiver with a unicast IP address). No zone will be created. The source node broadcasts the RREQ to the entire network, and the unicast receiver sends back the RREP and activates the nodes to join the forwarding group along the reverse route. After the unicast route is established, the source can send packets to the receiver until link breakage information is received, which will trigger route finding again (i.e., sending the RREQ and RREP again).

4.4 Multicast Ad Hoc On-Demand Distance Vector (MAODV) Routing Protocol

The MAODV routing protocol discovers multicast routes on demand using a broadcast route-discovery mechanism. A mobile node originates an RREQ message when it wishes to join a multicast group, or when it has data to send to a multicast group but it does not have a route to that group. Only a member of the desired multicast group may respond to a join RREQ.

If the RREQ is not a Join Request, any node with a fresh enough route (based on a group sequence number) to the multicast group may respond. If an intermediate node receives a join RREQ for a multicast group of which it is not a member, or if it receives an RREQ and it does not have a route to that group, it rebroadcasts the RREQ to its neighbors.

As the RREQ is broadcast across the network, nodes set up pointers to establish the reverse route in their route tables. A node receiving an RREQ first updates its route table to record the sequence number and the next-hop information for the source node. This reverse route entry may later be used to relay a response back to the

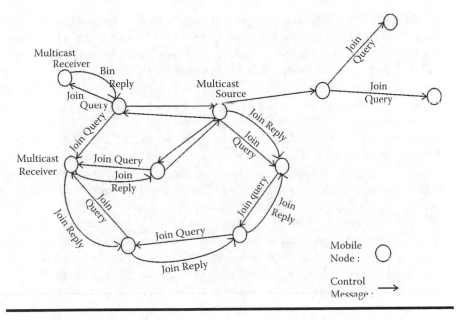

Figure 4.5 MAODV path discovery.

source. For join RREQs, an additional entry is added to the multicast route table. This entry is not activated unless the route is selected to be part of the multicast tree. If a node receives a join RREQ for a multicast group, it may reply if it is a member of the multicast group's tree and its recorded sequence number for the multicast group is at least as great as that contained in the RREQ. The responding node updates its route and multicast route tables by placing the requesting node's next-hop information in the tables, and then unicasts an RREP back to the source node. As nodes along the path to the source node receive the RREP, they add both a route table and a multicast route table entry for the node from which they received the RREP, thereby creating the forward path; see Figure 4.5 for a nonduplicate Join Request. It stores the upstream node ID (i.e., backward learning) and rebroadcasts the packet.

When the Join Request packet reaches a multicast receiver, the receiver creates or updates the source entry on its member. When a source node broadcasts an RREQ for a multicast group, it often receives more than one reply. The source node keeps the received route with the greatest sequence number and shortest hop count to the nearest member of the multicast tree for a specified period of time and disregards other routes. At the end of this period, it enables the selected next hop in its multicast route table, and unicasts an Activation Message (MACT) to this selected next hop. The next hop, on receiving this message, enables the entry for the source node in its multicast route table. If this node is a member of the multicast tree, it does not propagate the message any further. However, if this node is not a member of the multicast tree, it will have received one or more RREPs from its neighbors. It keeps the best next hop for its route to the multicast group, unicasts a MACT to that next hop, and enables the corresponding entry in its multicast route table. This process continues until the node that originated the RREP (member of the tree) is reached. The activation message ensures that the multicast tree does not have multiple paths to any tree node. Nodes only forward data packets along activated routes in their multicast route tables.

The first member of the multicast group becomes the leader for that group. The multicast group leader is responsible for maintaining the multicast group sequence number and broadcasting this number to the multicast group. This is done through a "Group Hello" message. The Group Hello contains extensions that indicate the multicast group's IP address and the sequence numbers (incremented with every Group Hello) of all multicast groups for which the node is the group leader. Nodes use the Group Hello information to update their request table.

Because AODV keeps hard state in its routing table, the protocol has to actively track and react to changes in this tree. If a member terminates its membership with the group, the multicast tree requires pruning. Links in the tree are monitored to detect link breakages. When a link breakage is detected, the node that is further from the multicast group leader (downstream of the break) is responsible for repairing the broken link. If the tree cannot be reconnected, a new leader for the disconnected downstream node is chosen as follows. If the node that initiated the route rebuilding is a multicast group member, it becomes the new multicast group leader. On the other hand, if it was not a group member and has only one next hop for the tree, it prunes itself from the tree by sending its next hop a prune message. This continues until a group member is reached.

Once separate partitions reconnect, a node eventually receives a Group Hello for the multicast group that contains group leader information that differs from the information it already has. If this node is a member of the multicast group, and if it is a member of the partition whose group leader has the lower IP address, it can initiate reconnection of the multicast tree.

MAODV uses a shared bidirectional multicast tree based on hard state, and any link breakages force actions to repair the tree. A multicast group leader maintains up-to-date multicast tree information by sending periodic Group Hello messages. Because MAODV unicasts the reply back to the source, if an intermediate node on the path moves away, the reply is lost and the route is lost. However, a broadcasted reply requires intermediate nodes not interested in the multicast group to drop the control packets, resulting in extra processing overhead. In MAODV, a potential multicast receiver must wait for a specified time, allowing for multiple replies to be received before sending an activation message along the multicast route that it selects.

4.5 Mesh-Based Routing Protocols

On-Demand Multicast Routing Protocol (ODMRP) is mesh-based and uses a forwarding group concept (only a subset of nodes forwards the multicast packets).

A soft-state approach is taken in ODMRP to maintain multicast group members. No explicit control message is required to leave the group.

In ODMRP, group membership and multicast routes are established and updated by the source on demand. When a multicast source has packets to send, but no route to the multicast group, it broadcasts a "Join Query" control packet to the entire network. This Join Query packet is periodically broadcast to refresh the membership information and update routes.

When an intermediate node receives the Join Query packet, it stores the source ID and the sequence number in its message cache to detect any potential duplicates. The routing table is updated with the appropriate node ID (i.e., backward learning) from which the message was received for the reverse path back to the source node. If the message is not a duplicate and the Time-to-Live (TTL) is greater than zero, it is rebroadcast. When the Join Query packet reaches a multicast receiver, it creates and broadcasts a "Join Reply" to its neighbors. When a node receives a Join Reply, it checks if the next-hop node ID of one of the entries matches its own ID. If it does, the node realizes that it is on the path to the source and thus is part of the forwarding group, and it sets the FG_FLAG (Forwarding Group Flag). It then broadcasts its own Join Table built upon matched entries.

The next-hop node ID field is filled by extracting information from its routing table. In this way, each forwarding group member propagates the Join Reply until it reaches the multicast source via the selected path (shortest). This whole process constructs (or updates) the routes from sources to receivers and builds a mesh of nodes, the forwarding group as shown in Figure 4.5.

After the forwarding group establishment and route construction process, sources can multicast packets to receivers via selected routes and forwarding groups. While it has data to send, the source periodically sends Join Query packets to refresh the forwarding group and routes. When receiving the multicast data packet, a node forwards it only when it is not a duplicate and the setting of the FG_FLAG for the multicast group has not expired. This procedure minimizes the traffic overhead and prevents sending packets through stale routes.

In ODMRP, no explicit control packets need to be sent to join or leave the group. If a multicast source wants to leave the group, it simply stops sending Join Query packets because it does not have any multicast data to send to the group. If a receiver no longer wants to receive data from a particular multicast group, it does not send the Join Reply for that group. Nodes in the forwarding group are demoted to non-forwarding nodes if not refreshed (no Join Tables received) before they time out.

In ODMRP, group membership and multicast routes are established and updated by the source on demand. Similar to on-demand unicast routing protocols, a request phase and a reply phase comprise the protocol, as shown in Figure 4.6.

While a multicast source has packets to send, it periodically broadcasts to the entire network a member advertising packet called a "Join Request." This periodic transmission refreshes the membership information and updates the route as follows. When a node receives a table, valid entries exist in the member table. "Join Tables"

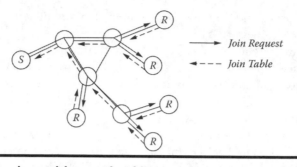

Figure 4.6   On-demand for membership setup and maintenance.

are broadcasted periodically to the neighbors. When a node receives a Join Table, it checks if the next node ID of one of the entries matches its own ID. If it does, the node realizes that it is on the path to the source and thus is part of the forwarding group. It then sets the FG-Flag and broadcasts its own Join Table built upon matched entries. The Join Table is thus propagated by each forwarding group member until it reaches the multicast source via the shortest path. This process constructs (or updates) the routes from sources to receivers and builds a mesh of nodes, the forwarding group. The visualized forwarding group concept is depicted in Figure 4.7.

The forwarding group is a set of nodes in charge of forwarding multicast packets. It supports the shortest paths between any member pairs. All nodes inside the "bubble" (multicast members and forwarding group nodes) forward multicast data packets.

Note that a multicast receiver can also be a forwarding group node if it is on the path between a multicast source and another receiver. The mesh provides richer connectivity among multicast members compared to trees. Flooding redundancy among the forwarding group helps overcome node displacements and channel fading.

4.5.1 Data Forwarding

After the group establishment and route construction process, a multicast source can transmit packets to receivers via selected routes and forwarding groups. Periodic control packets are sent only when outgoing data packets are still present.

When receiving a multicast data packet, a node forwards it only if it is not a duplicate and the setting of the FG-Flag for the multicast group has not expired. This procedure minimizes traffic overhead and prevents sending packets through stale routes.

4.5.2 Soft State

In ODMRP, no explicit control packets need to be sent to join or leave the group. If a multicast source wants to leave the group, it simply stops sending Join Request

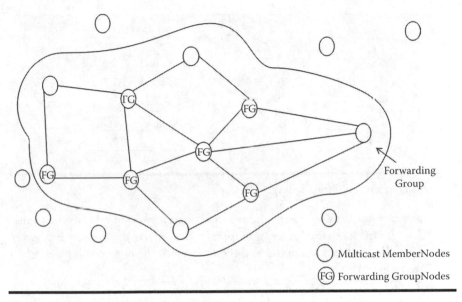

Figure 4.7   Forwarding group concept.

packets because it does not have any multicast data to send to the group. If a receiver no longer wants to receive data from a particular multicast group, it removes the corresponding entries from its Member Table and does not transmit the Join Table for that group. Nodes in the forwarding group are demoted to nonforwarding nodes if not refreshed (no Join Tables received) before they time out.

4.5.3 Data Structures

Network hosts running ODMRP are required to maintain the following data structures:

Member table: Each multicast receiver stores the source information in the member table. For each multicast group the node is participating in, the source ID and the time when the last Join Request is received from the source are recorded. If no Join Request is received from a source within the refresh period, that entry is removed from the member table.

Routing table: A routing table is created on demand and is maintained by each node. An entry is inserted or updated when a nonduplicate Join Request is received. The node stores the destination (i.e., the source of the Join Request) and the next hop to the destination (i.e., the last node that propagated the Join Request). The routing table provides the next-hop information when transmitting Join Tables (see Figure 4.8).

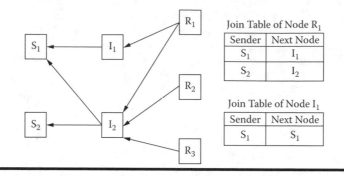

Figure 4.8 Join table forwarding.

Forwarding group table: When a node is a forwarding group node of the multi-cast group, it maintains the group information in the forwarding *group* table. The multicast group ID and the time when the node was last refreshed are recorded.

Message cache: The message cache is maintained by each node to detect duplicates. When a node receives a new Join Request or data, it stores the source ID and the sequence number of the packet. Note that entries in the message cache need not be maintained permanently. Schemes such as LRU (least recently used) or FIFO (First In–First Out) can be employed to expire and remove old entries and to prevent the size of the message cache from being expansive.

4.5.4 Unicast Capability

One of the major strengths of ODMRP is its unicast routing capability. Not only can ODMRP work with any unicast routing protocol, but it can also function as both multicast and unicast. Thus, ODMRP can run without any underlying unicast protocol.

The advantages of ODMRP are as follows:

- Low channel and storage overhead.
- Usage of up-to-date and shortest routes.
- Robustness to host mobility.
- Maintenance and exploitation of multiple redundant paths.
- Scalability to a large number of nodes.
- Exploitation of the broadcast nature of the wireless environment.
- Unicast routing capability.

4.6 Source Routing-Based Multicast Protocol (SRMP)

SRMP is an on-demand multicast routing protocol. This protocol is anchored on new idea-exploiting source and mesh routing as well as ad hoc features to provide robustness against mobility and to improve delay and throughput. It guarantees nodes' stability with respect to their neighbors, strong connectivity, availability of the used links, and higher battery lifetime. The path availability concept allows the protocol to distinguish between available and unavailable paths according to wireless link quality and nodes' stability.

The higher battery life concept biases the protocol toward choosing a link that tends to conserve power. The combination of these two criteria allows the selection of available and power-conserving links.

4.6.1 Protocol Overview

SRMP is a mesh-based multicast routing protocol. A mesh structure (an arbitrary subnetwork) is established on demand to connect group members, providing richer connectivity among multicast members. By building a mesh, packets can be efficiently delivered to multicast receivers in the case of node movements and topology changes. In addition, drawbacks of multicast trees can be avoided (e.g., intermittent connectivity, traffic concentration, frequent tree reconfiguration, and nonshortest path in a shared tree).

SRMP is based on a new source routing approach, in which the source route accumulates in the reply packet. The source routing concept is used by the DSR unicast protocol, allowing each data packet to carry in its header the list of nodes' addresses through which this packet must be transmitted.

During mesh establishment, SRMP uses the Forwarding Group (FG) nodes concept. The FG is a set of nodes responsible for forwarding multicast data between any member pairs. This scheme can be viewed as a "limited scope" flooding within a properly selected forwarding set. The key innovation of SRMP is to handle effective criteria in selecting FG nodes to achieve a compromise between the number of the selected nodes, the availability of the selected paths, and these paths' stability. Four metrics are considered to establish the mesh structure: association stability, link signal strength, link availability, and higher battery life.

4.6.2 Operation

A request phase and a reply phase comprise the protocol. The request phase invokes a route-discovery process to find routes to reach the multicast group. Different routes to the multicast group are set up during the reply phase through FG nodes selection and mesh construction. SRMP maintains four data structures: neighbor stability table, multicast message duplication table, multicast routing cache, and receiver multicast routing table.

The request phase starts when a source node, which is not a group member, wishes to join the group. It invokes a route-discovery procedure toward the multicast group through broadcasting a Join Request packet to neighbors. To eliminate the possibility of receiving multiple copies of Join Request, each node detects the duplication in reception and drops it. A reply phase starts at each multicast receiver receiving a Join Request. It first checks for stability among its neighbors based on the four selection metrics mentioned previously. A neighbor is selected as an FG node, if these metrics satisfy predefined thresholds. The receiver starts sending Join Reply messages to the selected neighbors, setting their types as member nodes in its neighbor stability table. Each neighbor, receiving a Join Reply, creates an entry in its multicast routing cache for the multicast group, setting its state as FG node and storing the reversed accumulated route in the received Join Reply. The source route from the multicast receiver to the requesting node is accumulated in a route record field in each Join Reply packet. The process continues until it reaches the source of the request, constructing a mesh of FG nodes connecting group members. After establishment of the group, the source floods its data packets along the FG mesh via the routes stored in its multicast routing cache toward the multicast receivers.

Mesh refreshment in SRMP is a simple mechanism; it requires no extra control overhead. The multicast source refreshes the corresponding route entries in its multicast routing cache for each data packet it transmits to the multicast group. Each FG node forwarding this packet scans the packet header for the used route, refreshing the corresponding route entry in the multicast routing cache. In addition, a multicast receiver scans the header of each received data packet, refreshing the corresponding entry in the receiver multicast routing table.

SRMP reacts to node mobility on demand, such that it detects link failure during data transmission through the use of MAC layer support. Two mechanisms are addressed: (1) link repair between two FG nodes, and (3) link repair between a multicast receiver and an FG node, making use of a Multicast RERR message. In addition, SRMP employs an effective pruning mechanism allowing a member node to leave the multicast session. It deals with two cases: FG node pruning and multicast receiver pruning. A multicast source revokes its member status simply through stopping data transmission and removing entries concerning the group from its multicast routing cache. A multicast receiver may prune itself by sending a Leave Group message to all member neighbors, and deleting from its receiver table all corresponding entries.

4.7 Multicasting with Quality-of-Service (QoS) Guarantees

Most conventional multicast protocols were designed for best-effort data. They construct multicast trees primarily based on connectivity. Such trees may be unsatisfactory due to the lack of resources. Some algorithms provide heuristic solutions to the Nondeterministic Polynomial (NP) time–complete constrained Steiner tree problem, which is to find the delay-constrained, least-cost multicast trees. These algorithms, however, are not practical in the Internet environment because they have excessive computation overhead, require knowledge about the global network state, and do not handle dynamic group membership. Jia's distributed algorithm does not compute any path or assume the unicast routing table can provide it. However, this algorithm requires excessive message processing overhead. The Spanning Join Protocol handles dynamic membership and does not require any global network state. However, it has excessive communication and message-processing overhead because it relies on full flooding to find a feasible tree branch to connect a new member. Model Integrated Computing (MIC), proposed by Faloutsos et al., alleviates but does not eliminate the flooding behavior. In addition, an extra control element, called "manager router," is introduced to handle the Join Requests of new members.

QoS routing not only requires finding a route from a source to a destination, but also the route must satisfy the end-to-end QoS requirement, often given in terms of bandwidth or delay. Quality of service is more difficult to guarantee in ad hoc networks than in other types of networks, because the wireless bandwidth is shared among adjacent nodes and the network topology changes as the nodes move. This requires extensive collaboration between the nodes, both to establish the route and to secure the resources necessary to provide the QoS. The important QoS hierarchical multicast routing protocol for mobile ad hoc networks is Hierarchical QoS Multicast Routing Protocol (HQMRP). It not only ensures fast convergence but also provides multiple guarantees for satisfying multiple constraints. HQMRP also allows that an ad hoc group member can join and leave the multicast group dynamically.

4.8 Energy-Efficient Multicast Routing Protocols

A multicast packet is delivered to multiple receivers along a network structure such as a tree or mesh, which is constructed once a multicast group is formed. However, the network structure is fragile due to node mobility, and thus some members may not be able to receive the multicast packet. To improve the packet delivery ratio, multicast protocols in MANETs usually employ control packets to periodically refresh the network structure. Mesh-based protocols are more robust to mobility than tree-based protocols due to many redundant paths between mobile nodes in the mesh. However, a multicast mesh may perform worse in terms of energy efficiency because it uses costlier broadcast-style communication involving forward nodes than multicast trees. Another important aspect of energy efficiency is balanced energy consumption among all participating mobile nodes. To maximize the lifetime of a MANET, care has to be taken not to unfairly burden any particular node with many packet delivery operations.

A new multicast routing protocol, the Two-Tree Multicast (TTM) Protocol for MANETs, is discussed here. By maintaining two trees, called primary and alternative trees, TTM consumes less energy than the mesh-based multicast and performs better than the conventional tree-based multicast in terms of its packet delivery ratio.

TTM is a tree-based multicast protocol employing multidestined unicast-based trees and thus consumes less energy than mesh-based protocols. It uses shared trees rather than per-source trees to avoid tree construction and maintenance overhead. Unique to TTM is the use of two trees called primary and alternative trees for a multicast group. When the primary tree is unusable or overloaded, the alternative tree takes on the responsibility of the primary tree, and a new alternative tree is immediately constructed for future use. A group senses with the larger remaining energy. Larger battery energy is selected as the root node of the new alternative tree. Two trees can reduce to latest problem when a link error occurs on the primary tree by immediately switching to the alternative tree. Tree replacement is also for alternatives the energy balance problem in a shared-tree multicast.

4.9 Application-Dependent Multicast Routing

A prominent application of mobile ad hoc networks is direct wireless communication between vehicles caught in a traffic jam or for a vehicle involved in an accident; in these applications, the vehicles are equipped with a computer-controlled radio modem allowing them to contact other equipped vehicles in their vicinity. Adhering to the abstract definition of an ad hoc network, we assume no fixed infrastructure to support the communication. The best applications of intervehicle communication are to provide improved comfort and additional safely in driving. Our aim is to make these applications feasible by enabling the dissemination of information among participating vehicles. As an example application, an equipped vehicle identifies itself as crashed by vehicular sensors that detect events like airbag ignition, and it can report the accident instantly to equipped vehicles nearby. An algorithm is presented here to disseminate such a message among the other equipped vehicles on the road. Multihopping allows us to enlarge the area in which an equipped vehicle receives this message. If the message reaches a vehicle for which the warning is relevant, the driver can be informed early by the system.

4.9.1 Role-Based Multicast Routing Protocol

Many studies in ad hoc networking propose mobility patterns in two-dimensions. The hosts change their speed and direction more or less randomly follow the road, which allows us to reduce mobility to one dimension. As an essential requirement in networking, every host vehicle possesses a unique identifier, but the set of existing identifiers can easily exceed a practical size of fixed host or server tables. In addition, newly manufactured vehicles equipped with the system join the set of existing identifiers, whereas the identifiers, as an ad hoc network formed on the road, will only connect a small subset of identifiers. In applications of intervehicle communication, a vehicle often needs to address other nearby vehicles whose identities are a priori unknown rather then send a packet to specifically identified vehicles.

In the scenario with a road accident, the crashed vehicle wants to inform the other vehicles that are approaching the hazardous area. To make the system work, the vehicles need to be aware of their locations; many vehicles can utilize navigational systems like Global Positioning Systems (GPS).

However, in this application aspects other than location determine whether a vehicle belongs to the multicast group. Taking the driving direction into account, a vehicle can distinguish more reliably whether it is approaching the dangerous spot or not. Also, if it employs a digital road map, the vehicle improves its ability to classify the situation on divided highways, and usually does not accidentally harm vehicles in the other driving direction. Finally, the velocity of a vehicle puts an individual dead line a message delivery for each potential recipient inside the multicast region. When the vehicle receives the warning after it has passed the accident, the message is useless to the driver due to the high speed of mobile nodes. The ultimate point where a vehicle has to be informed is within the braking distance of the accident. An equation is used to calculate each vehicle's braking distance depending on its velocity. Hence, the multicast group is further limited to those vehicles inside the multicast region that are still able to stop in front of the accident.

Assuming that warning about an accident is a relatively rare event, it seems suitable to design the protocol to work on demand. However, a warning message is not much bigger than a control packet sent during a request phase. Hence, the response phase is avoided group and simply apply flooding to reach the multicast group, a packet sent by one host can reach multiple receivers simultaneously.

Each vehicle maintains the set of neighbors. It constantly updates N according to the notification from the data-link layer; also, each vehicle associates the sets with the warning message. Every time the system receives the message from a sender, it adds sender identity to S on the first reception of the message. S is initialized with corresponding sender identity, and the system switches into one of two states: "Wait to Respond" or "Wait" for neighbor.

If $N/S \neq 0$, the system has a neighbor other than the sender of the personally received message; it enters the "Wait to Respond" mode. The message header contains the position of its sender by knowing its own position; the system determines waiting time (WT) depending on the distance to the accident on the side of the highway where the accident happened.

Unlike the traditional multicast schemes, here the multicast group is implicitly defined as the set of nodes within a specified area. This specified area is returned as the "multicast region," and the set of nodes in the multicast region as the location-based multicast group. If a host resides within the multicast group at a given time, it will automatically become a member of the corresponding multicast group at that time. A location-based multicast group may be used for sending a message that is likely to be of interest to everyone in a specified area.

4.9.2 Location-Based Multicast Protocol

Two approaches may be used to implement a location-based multicast:

- Maintain a multicast tree, such that all nodes within the multicast region at any time belong to the multicast tree. The tree would need to be updated wherever nodes enter or leave the multicast region.
- Do not mountain a multicast tree; in this case, multicasts may be performed using some sort of "flooding" schema.

Using the flooding algorithm, provided that intended multicast group sensors are reachable from the sender, the sensors should eventually receive a multicast message; it is possible that some group sensors will not receive the packet (e.g., when they are unreachable from the sender or multicast messages are lost due to the transmission errors).

This algorithm would be very simple and robust but would not be very efficient. Using the location information of the source and the specified multicast region in the following algorithm, it tries to reduce the number of nodes outside the multi-cast region to whom a multicast packet is propagated.

4.9.2.1 Location-Based Multicast Algorithm

This approach makes use of location-based multicast groups and utilizes the location information to reduce the multicast delivery; overhead location information is provided by a Global Positioning System (GPS), with the host knowing its physical location.

4.9.2.2 Multicast Region and Forwarding Zone

In regard to multicast regions, consider a node S that needs to multicast a message to all nodes that are currently located within a certain geographical region. This specific area is called the "multicast region."

4.10 Summary

The challenges faced by multicast routing protocols for ad hoc wireless networks are much more complex than those faced by their wired network counterparts. In this chapter, the problem of multicast routing in ad hoc wireless networks was studied. After identifying the main issues involved in the design of a multicast routing protocol, a classification of the existing multicasting protocols was given. Several of these multicast routing protocols were described in detail with suitable examples. The advantages and drawbacks involved in each protocol were also listed. Some energy-conserving multicasting routing protocols were presented, as most of the nodes in ad hoc wireless networks are battery operated.

Reliable multicasting has become indispensable for the successful deployment of ad hoc wireless networks as these networks support important applications such as military battlefields and emergency operations. Real-time multicasting that supports bounded delay delivery for streaming data is also a potential avenue for research. Security is another necessary requirement, which is still lacking in ad hoc multicast routing protocols, as multicast traffic of important and high-security (e.g., military) applications may pass through unprotected network components (routers or links). Thus, multicasting in ad hoc networks is a significant problem that merits further exploration.

4.11 Problems

4.1 Comment on the scaling properties of source-initiated and receiver-initiated multicast protocols with respect to the number of sources and receivers in the group. Which of them would be suitable for (1) a teacher multicasting his or her lectures to a set of students (assume the students do not interact with one another), and (2) a distributed file-sharing system?

4.2 Is hop length always the best metric for choosing paths? In an ad hoc network with a number of nodes, each differing in mobility, load generation characteristics, interference level, and so forth, what other metrics are possible?

4.3 Link-level broadcast capability is assumed in many of the multicast routing protocols. Are such broadcasts reliable? Give some techniques that could be used to improve the reliability of broadcasts.

4.4 What are the two basic approaches for maintenance of the multicast tree in bandwidth-efficient multicast protocol (BEMRP)? Which of the two performs better? Why?

4.5 Calculate the efficiency of the ODMRP and DCMP protocols from Figure 4.6.

4.6 What makes CAMP an efficient protocol?

4.7 What are the two different topology maintenance approaches? Which of the two approaches is better when the topology is highly dynamic? Give reasons for your choice.

4.8 Comment on how content-based multicasting (CBM) could be advantageous or disadvantageous as far as the bandwidth utilization of the network is concerned.

Bibliography

R. Bagrodia, M. Gerla, J. Hsu, W. Su, and S-J. Lee, A performance comparison study of ad hoc wireless multicast protocols, Proceedings of the 19th Annual Joint Conference of the IEEE Computer and Communications Societies, March 2000, pp. 565–574.

F. Bai, N. Sadagopan, and A. Helmy, Important: A framework to systematically analyze the impact of mobility on performance of routing protocols for ad hoc networks, in Proceedings of the IEEE Information Communications Conference (INFOCOM 2003), San Francisco, April 2003.

E. Bommaiah, M. Liu, A. McAuley, and R. Talpade, AM route: Adhoc multicast routing protocol, Aug. 1998, work in progress.

J. Broch, D. A. Maltz, D. B. Johnson, Y-C. Hu, and J. Jetcheva, A performance comparison of multi-hop wireless ad hoc network routing protocols, in Proceedings of ACM/IEEE MOBICOM'98, Dallas, TX, Oct. 1998, pp. 85–97.

E. Cheng, On-demand multicast routing in mobile ad hoc networks, M. Eng. thesis, Carleton University, 2001.

S. Corson and J. Macker, Mobile ad hoc networking (MANET): Routing protocol performance issues and evaluation considerations, RFC 2501, Jan. 1999, http://www.ietf.org/rfc/rfc2501.txt.

S. E. Deering and D. R. Cheriton, Multicast routing in datagram inter-networks and extended LANs, *ACM Transactions on Computer Systems*, 8(2):85–110, May 1990.

J. J. Garcia-Luna-Aceves and E. L. Madruga, A multicast routing protocol for ad-hoc networks, in Proceedings of IEEE INFOCOM'99, New York, March 1999, pp. 784–792.

P. Johansson, T. Larsson, N. Hedinan, and B. Mielczarek, Routing protocols for mobile ad-hoc networks: A comparative performance analysis, in Proceedings of ACM/IEEE MOBICOM'99, Seattle, WA, Aug. 1999.

S. Lee, M. Gerla, and C. Chiang, On-demand multicast routing protocol, Proceedings of IEEE Wireless Communications and Networking Conference (WCNC'99), pp. 1298–1302, 1999.

The implementation of ODMRP, http://www.monarch.cs.rice.edu/multicast_extensions.html.

L. Layuan and L. Chunlin, A QoS-guaranteed multicast routing protocol, *Computer Communications*, 27(1):59–69, 2004.

S-L. Lee, Routing and multicast strategies in wireless mobile ad hoc networks, Ph.D. thesis, University of California, 2000.

S-J. Lee, W. Su, J. Hsu, M. Gerla, and R. Bagrodia, A performance comparison study of ad hoc wireless multicast protocols, Proceedings of the IEEE INFOCOM 2000, Vol. 2, March 2000, pp. 565–574.

H. Lim and C. Kim, Multicast tree construction and flooding in wireless ad hoc networks, Proceedings of the 3rd ACM International Workshop on Modeling, Analysis and Simulation of Wireless and Mobile Systems, Aug. 2000, pp. 61–68.

K. Ohfiiczka and C. Tsudik, Multicast routing issues in ad hoc networks, IEEE International Conference on Universal Personal Communication (ICUPC'98), Oct. 1998.

E. Royer and C. E. Perkins, Multicast operation of the ad-hoc on-demand distance vector routing protocol, Proceedings of the 5th ACM/IEEE Annual Conference on Mobile Computing and Networking, Aug. 1999, pp. 207–218.

S. Singh, C. S. Raghavendra, and J. Stepanek, Power aware broadcasting in mobile ad hoc networks, technical report, Oregon State University, 1999.

J. E. Wieselthier, G. D. Nguyen, and A. Ephremides, Algorithms for energy-efficient multicasting in ad hoc wireless networks, Proceedings of Military Communication Conference (MILCOM 1999), Vol. 2, pp. 1414–1418, Nov. 1999.

H. Woesner, J. Ebert, M. Schlager, and A. Wolisz, Power-saving mechanisms in emerging standards for wireless LANs: The MAC level perspective, *IEEE Personal Communications*, 5(3):40–48, June 1998.

H. Y. Youn, C. Yu, B. Lee, and S. Moh, Energy efficient multicast in ad hoc networks, in *Handbook of ad hoc wireless networks*, CRC Press, Boca Raton, FL, 2002.

Chapter 5

Transport Protocols for Ad Hoc Networks

5.1 Introduction

Ad hoc networks are complex distributed systems that consist of wireless mobile or static nodes that can freely and dynamically self-organize. In this way they form arbitrary and temporary "ad hoc" network topologies, allowing devices to seamlessly interconnect in areas with no preexisting infrastructure. Recently, the introduction of new protocols such as Bluetooth, IEEE 802.11, and HyperLAN are making possible the deployment of ad hoc networks for commercial purposes. As a result, considerable research efforts have been put into this new challenging wireless environment.

Transmission Control Protocol (TCP) was designed to provide reliable end-to-end delivery of data over unreliable networks. In theory, TCP should be independent of the technology of the underlying infrastructure. In particular, TCP should not care whether the Internet Protocol (IP) is running over wired or wireless connections. In practice, it does matter because most TCP deployments have been carefully designed based on assumptions that are specific to wired networks. Ignoring the properties of wireless transmission can lead to TCP implementations with poor performance.

In ad hoc networks, the principal problem of TCP lies in performing congestion control in case of losses that are not induced by network congestion. Because bit error rates are very low in wired networks, nearly all TCP versions nowadays assume that packet losses are due to congestion. Consequently, when a packet is detected to be lost, either by timeout or by multiple duplicated acknowledgment (ACK) packets, TCP slows down the sending rate by adjusting its congestion window. Unfortunately, wireless networks suffer from several types of losses that are not related to congestion, making TCP not adapted to this environment. Ad hoc networks inherit several features like high bit error rates and path asymmetry, and add new problems that come from mobility and multihop communications, such as network partitions, route failures, and hidden (or exposed) terminals.

5.2 Transmission Control Protocol's (TCP's) Challenges and Design Issues in Ad Hoc Networks

The performance of TCP degrades in ad hoc networks. This is because TCP has to face new challenges due to several reasons specific to these networks: lossy channels, hidden and exposed stations, path asymmetry, network partitions, route failures, and power constraints.

5.2.1 Challenges

5.2.1.1 Lossy Channels

The main causes of errors in wireless channels are the following:

Signal attenuation: This is due to a decrease in the intensity of the electromagnetic energy at the receiver (e.g., due to long distance), which leads to a low Signal-to-Noise Ratio (SNR).

Doppler shift: This is due to the relative velocities of the transmitter and the receiver. Doppler shift causes frequency shifts in the arriving signal, thereby complicating the successful reception of the signal.

Multipath fading: Electromagnetic waves reflecting off objects or diffracting around objects can result in the signal traveling over multiple paths from the transmitter to the receiver. Multipath propagation can lead to fluctuations in the amplitude, phase, and geographical angle of the signal received at a receiver.

To increase the success of transmissions, link-layer protocols implement the following techniques: Automatic Repeat Request (ARQ), Forward Error Correction (FEC), or both. For example, IEEE 802.11 implements ARQ, so when a transmitter detects an error, it will retransmit the frame, and error detection is timer based. Bluetooth implements both ARQ and FEC on some synchronous and asynchronous connections. Note that packets transmitted over a fading channel may cause a routing protocol to incorrectly conclude that there is a new one-hop neighbor. This

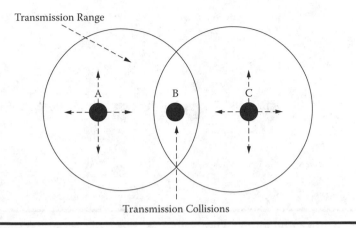

Figure 5.1 Hidden terminal problem.

one-hop neighbor could provide a shorter route to even more distant nodes. Unfortunately, this new shorter route is usually unreliable. But some routing protocols, like Destination-Sequenced Distance Vector (DSDV) and Ad Hoc On-Demand Distance Vector (AODV) in a real network, find that neither of these protocols can provide a stable multihop route because of the physical channel behaviors and especially fading.

5.2.1.2 Hidden and Exposed Stations

In ad hoc networks, stations may rely on a physical carrier-sensing mechanism to determine an idle channel, such as in the IEEE 802.11 Distributed Coordination Function (DCF). This sensing mechanism does not solve completely the hidden station and the exposed station problems. Before explaining these problems, we need to clarify the term "transmission range." The transmission range is the range, with respect to the transmitting station, within which a transmitted packet can be successfully received.

A typical hidden terminal situation is depicted in Figure 5.1. Stations A and C have a frame to transmit to station B. Station A cannot detect C's transmission because it is outside the transmission range of C. Station C (respect to A) is therefore "hidden" to station A (respect to C). Because A's and C's transmission areas are not disjoint, there will be packet collisions at B. These collisions make the transmission from A and C toward B problematic. To alleviate the hidden station problem, virtual carrier sensing has been introduced. It is based on a two-way handshaking that precedes data transmission. Specifically, the source station transmits a short control frame, called Request-to-Send (RTS), to the destination station. Upon receiving the RTS frame, the destination station replies by a Clear-to-Send (CTS) frame, indicating that it is ready to receive the data frame. Both RTS and

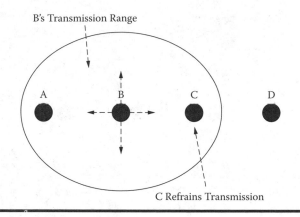

Figure 5.2 Exposed terminal problem.

CTS frames contain the total duration of the data transmission. All stations receiving either RTS or CTS will keep silent during the data transmission period (e.g., station C in Figure 5.1).

However, the hidden station problem may persist in IEEE 802.11 ad hoc networks even with the use of the RTS–CTS handshake. This is due to the fact that the power needed for interrupting a packet reception is much lower than that for delivering a packet successfully. In other words, a node's transmission range is smaller than the sensing node range.

The exposed station problem results from a situation where a transmission has to be delayed because of the transmission between two other stations within the sender's transmission range. In Figure 5.2, consider a typical scenario where the exposed terminal problem occurs. Let us assume that A and C are within B's transmission range, and A is outside C's transmission range. Let us also assume that B is transmitting to A, and C has a frame to be transmitted to D. According to the carrier-sensing mechanism, C senses a busy channel because of B's transmission. Therefore, station C will refrain from transmitting to D, although this transmission would not cause interference at A. The exposed station problem may thus result in a reduction of channel utilization.

It is worth noting that hidden terminal and exposed terminal problems are correlated with the transmission range. By increasing the transmission range, the hidden terminal problem occurs less frequently. On the other hand, the exposed terminal problem becomes more important as the transmission range identifies the area affected by a single transmission.

5.2.1.3 Path Asymmetry

Path asymmetry in ad hoc networks may appear in several forms like bandwidth asymmetry, loss rate asymmetry, and route asymmetry.

Bandwidth asymmetry: Satellite networks suffer from high bandwidth asymmetry, resulting from various engineering trade-offs (such as power, mass, and volume) as well as the fact that for scientific space missions, most of the data originates at the satellite and flows to the earth. The return link is not used, in general, for data transferring. For example, in broadcast satellite networks the ratio of the bandwidth of the satellite–earth link over the bandwidth of the earth–satellite link is about 1,000. On the other hand, in ad hoc networks, the degree of bandwidth asymmetry is not very high. For example, the bandwidth ratio lies between 2 and 54 in ad hoc networks that implement the IEEE 802.11 (version g) protocol. The asymmetry results from the use of different transmission rates. Because of these different transmission rates, even symmetric source destination paths may suffer from bandwidth asymmetry.

Loss rate asymmetry: This type of asymmetry takes place when the backward path is significantly lossier than the forward path. In ad hoc networks, this asymmetry is due to the fact that packet losses depend on local constraints that can vary from place to place. Note that loss rate asymmetry may produce bandwidth asymmetry. For example, in multirate IEEE 802.11 protocol versions, senders may use the Auto-Rate-Fallback (ARF) algorithm for transmission rate selection. With ARF, senders attempt to use higher transmission rates after consecutive transmission successes, and revert to lower rates after failures. So, as the loss rate increases, the sender will keep using low transmission rates.

Route asymmetry: Unlike the previous two forms of asymmetry, where the forward path and the backward path can be the same, route asymmetry implies that distinct paths are used for TCP data and TCP ACKs. This asymmetry may be an artifact of the routing protocol used. Route asymmetry increases routing overheads and packet losses in cases of high degrees of mobility, because when nodes move, using distinct forward and reverse routes increases the probability of route failures experienced by TCP connections. However, this is not the case with static networks or networks that have a low degree of mobility, like the case of a network with routes of high lifetime compared to the session transfer time. So, it is up to the routing protocols to select symmetric paths when such routes are available in the case of ad hoc networks of high mobility.

5.2.1.4 Network Partition

An ad hoc network can be represented by a simple graph G. Mobile stations are the "vertices." A successful transmission between two stations is an undirected "edge." Network partition happens when G is disconnected. The main reason of this disconnection in MANETs is node mobility. Another factor that can lead to network partition is the energy-constrained operation of nodes. An example of network partition is illustrated in Figure 5.3. In this figure, dashed lines are the links between

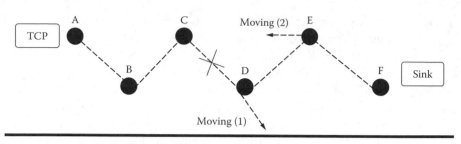

Figure 5.3   Network partition scenario.

nodes. When node D moves away from node C, this results in a partition of the network into two separate components. Clearly, the TCP agent of node A cannot receive the TCP ACK transmitted by node F. If the disconnectivity persists for a duration greater than the Retransmission TimeOut (RTO) of node A, the TCP agent will trigger the exponential backoff algorithm, which consists of doubling the RTO whenever the timeout expires. Originally, TCP does not have an indication about the exact time of network reconnection. This lack of indication may lead to long idle periods during which the network is connected again, but TCP is still in the backoff state.

5.2.1.5   Routing Failures

In wired networks, route failures occur very rarely. In MANETs, they are frequent events. The main cause of route failures is node mobility. Another factor that can lead to route failures is the link failures due to the contention on the wireless channel, which is the main cause of TCP performance degradation in MANETs. The route reestablishment duration after route failure in ad hoc networks depends on the underlying routing protocol, mobility pattern of mobile nodes, and traffic characteristics. If a TCP sender does not have indications on the route reestablishment event, the throughput and session delay will degrade because of the large idle time. Also, if the new route established is longer or shorter in term of hops, then the old-route TCP will face a brutal fluctuation in Round-Trip Time (RTT).

In addition, ad hoc networks, routing protocols that rely on broadcast Hello messages to detect neighbors' reachability, may suffer from the problem of "communication gray zones." In these zones, data messages cannot be exchanged, although broadcast Hello messages and control frames indicate that neighbors are reachable. So when sending data messages, routing protocols will experience routing failures.

5.2.1.6   Power Constraints

Because batteries carried by each mobile node have a limited power supply, processing power is limited. This is a major issue in ad hoc networks, as each node is acting as an end system and a router at the same time, with the implication that additional energy is required to forward and relay packets. TCP must use this scarce power resource in an "efficient" manner. Here, "efficiency" means minimizing the number of unnecessary retransmissions at the transport layer as well as at the link layer. In general, in ad hoc networks there are two correlated power problems: the first one is "power saving" that aims at reducing the power consumption, and the second one is "power control" that aims at adjusting the transmission power of mobile nodes. Power-saving strategies have been investigated at several levels of a mobile device, including physical layer transmissions, operation systems, and applications. Power control can be jointly used with routing or transport agents to improve the performance of ad hoc networks. Power constraint communications reveal also the problem of cooperation between nodes, as nodes may not participate in routing and forwarding procedures to save battery power.

5.2.2 Design Goals

The following goals are to be met when designing the transport-layer protocol for ad hoc networks:

- The throughput for a connection should be maximized.
- The throughput fairness should be provided.
- Connection setup time should be minimized.
- Connection maintenance overhead should be minimized.
- Should incorporate mechanisms for congestion and flow control.
- Should provide both reliable and unreliable transport.
- Should use the available bandwidth efficiently.
- Should be aware of resource constraints, e.g., power and buffer size.
- To improve performance, lower layer information should be efficiently used.
- The cross-layer interactions should be efficient, scalable, and protocol independent.

5.3 TCP Performance over That of Mobile Ad Hoc Networks (MANETs)

5.3.1 TCP Performance

To implement a mobile TCP, the following factors have to be considered.

5.3.1.1 Noncongestion Delay

One of the main problems that TCP has over MANETs is that it assigns all packet losses to congestion, and that is controlled by the congestion control scheme. The congestion window is reduced, and resets the retransmission timer to a backoff interval. Congestion control algorithms need to be used only in the event of genuine network congestion. Note that Jacobson assumes that packet loss due to damage in transit is rare; hence, most probably packets get lost due to network congestion and not due to damage. In Jacobson, it has been stated that the congestion control scheme is insensitive to damage loss. High loss rates due to damage of one packet per window (e.g., 12 to 15 percent for an 8-packet window) degrade TCP throughput by 60 percent. The additional degradation from the congestion avoidance window shrinking escalates the problem.

5.3.1.2 Serial Timeouts

Frequent disconnections cause a condition called "serial timeouts" at the TCP sender. This happens when the retransmission timer at the sender is doubled with each unsuccessful retransmission attempt, to reduce the transmission rate. Thus, when the mobile is reconnected, TCP will take a long time to recover from such a reduction and data will not be transmitted for a period of time.

5.3.1.3 Packet Size Variation

Packet size over wireless links is typically much smaller than packet size over wired links. As a result, each packet on the wired networks gets fragmented when transmitted over the wireless link. Therefore, finding the optimal packet size on the wireless link is a key issue for performance.

5.3.1.4 The Data and Acknowledgment (ACK)
Packet Collision Problem

The collision avoidance mechanism of IEEE 802.11b does not prevent all collisions. Because TCP traffic is bidirectional (with data packets in one direction and ACK packets in the opposite direction), there can be TCP data packet and ACK packet collisions. These collisions cause Medium Access Control (MAC)-layer retransmissions or TCP-layer retransmissions when link-layer error recovery is not used. Here, Jacobson tested the retransmission rate for User Datagram Protocol (UDP) and TCP within an environment unlikely to have interference. In UDP, the retransmissions were relatively slow (close to 1 percent), but when they used TCP, the retransmissions rose to 5 percent. They blamed the increase in the retransmission rate to TCP data–ACK packet collisions. The reduction of performance is not important, but the performance is even lower if MAC-layer recovery is not used. The TCP throughput without MAC-layer retransmission is 23 percent lower than that with MAC-layer retransmission. The UDP-streaming performance, even without MAC-layer retransmission, is slighter higher than that of TCP with MAC-layer retransmission.

5.3.2 Other Problems

The other problems in the lower layers are outlined below.

5.3.2.1 Spread of Stale Routes

When the nodes move, routes change, and they have to be updated as soon as possible. Even in slowly changing topologies, the TCP sender is very slow in purging stale routes from its cache, resulting in repeated rooting failures. After these failures, the intermediate nodes reply to the route request with routes in their caches, complicating this problem as they reply sometimes with stale routes. That gets worse when other nodes overhear the stale routes in the replies; therefore, spare routes are spread over the network, producing more routing failures. Those are some of the facts that have a detrimental impact on TCP performance. It can be solved by dynamically adjusting the route cache timeout depending on the observed route failure rate.

5.3.2.2 The Medium Access Control (MAC)-Layer Rate Adaptation Problem

This concerns the MAC-layer rate adaptation algorithm. The MAC-layer rate adaptation is supposed to increase the throughput when the channel error rate is high. A poor rate adaptation algorithm could decrease throughput. The Multiplicative Increase–Multiplicative Decrease (MIMD) rate adaptation algorithm causes the periodic TCP packet retransmissions. This "bandwidth-probing" mechanism causes "network thrashing" on the wireless local area network (WLAN). The wireless channel resource is wasted on many MAC-layer retransmissions, and many efficient or inefficient TCP-layer retransmissions occur. Rather, a better rate adaptation algorithm is required. A new "bandwidth-probing" algorithm will be needed.

The cause of TCP performance degradation in MANETs is due to four major problems:

- TCP is unable to distinguish between losses due to route failures and network congestion.
- TCP suffers from frequent route failures.
- The contention on wireless channel.
- TCP unfairness.

5.4 Ad Hoc Transport Protocols

5.4.1 Split Approaches

This section addresses the problem of frequent route failures in MANETs. Figure 5.4 shows the classification of transport-layer protocols.

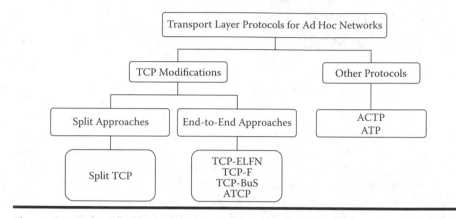

Figure 5.4   Classification of transport-layer protocols.

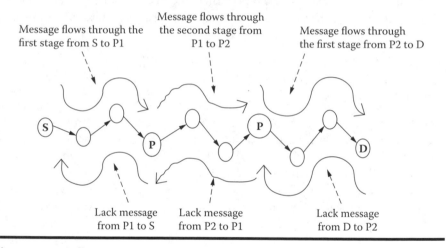

Figure 5.5   Split TCP.

5.4.1.1   Split TCP

TCP connections that have a large number of hops suffer from frequent route failures due to mobility. To improve the throughput of these connections and to resolve the unfairness problem, the Split TCP scheme was introduced to split long TCP connections into shorter localized segments (see Figure 5.5). The interfacing node between two localized segments is called a "proxy."

The routing agent decides if its node has the role of proxy according to the interproxy distance parameter. The proxy intercepts TCP packets, buffers them, and acknowledges their receipt to the source (or previous proxy) by sending a Local Acknowledgment (LACK). Also, a proxy is responsible for delivering the packets, at an appropriate rate, to the next local segment. Upon the receipt of a LACK (from the next proxy or from the final destination), a proxy will purge the packet from its buffer. To ensure source-to-destination reliability, an ACK is sent by the destination to the source similarly to the standard TCP. In fact, this scheme also splits the transport-layer functionalities into end-to-end reliability and congestion control. This is done by using two transmission windows at the source which are the congestion window and the end-to-end window. The congestion window is a subwindow of the end-to-end window. Although the congestion window changes in accordance with the rate of arrival of LACKs from the next proxy, the end-to-end window will change in accordance with the rate of arrival of the end-to-end ACKs from the destination. At each proxy, there would be a congestion window that would govern the rate of sending between proxies.

5.4.1.1.1 Advantages

An interproxy distance of between 3 and 5 has a good impact on both throughput and fairness. An improvement of up to 30 percent can be achieved in the total throughput by using Split TCP.

5.4.1.1.2 Disadvantages

The drawbacks are large buffers and network overheads. Also, this makes the role of proxy nodes more complex, as for each TCP session they have to control packet delivery to succeeding proxies.

5.4.2 End-to-End Approaches

This addresses the problem of TCP's inability to distinguish between losses due to route failures and those due to network congestion in MANETs.

5.4.2.1 TCP Feedback (TCP-F)

In mobile ad hoc networks, topology may change rapidly due to the movement of Mobile Hosts (MHs). The frequent topology changes result in sudden packet losses and delays. TCP misinterprets such losses as congestion and invokes congestion control, leading to unnecessary retransmission and loss of throughput. To overcome this problem, TCP feedback (TCP-F) was proposed so that the sender can distinguish between route failure and network congestion. In this scheme, the sender is forced to stop transmission without reducing window size upon route failure. As soon as the connection is reestablished, fast retransmission is enabled.

TCP-F relies on the network layer at an intermediate node to detect the route failure due to the mobility of its downstream neighbor along the route. A sender can be in an "active state" or a "snooze state." In the active state, the transport

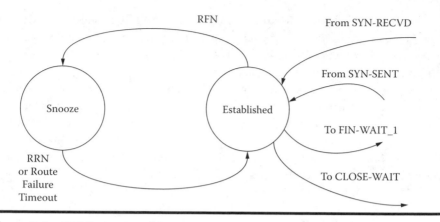

Figure 5.6 TCP-F state machine.

layer is controlled by the normal TCP. As soon as an intermediate node detects a broken route, it explicitly sends a Route Failure Notification (RFN) packet to the sender and records this event. Upon reception of the RFN, the sender goes into the snooze state, in which the sender completely stops sending further packets and freezes all of its timers and the values of state variables such as RTO and congestion window size. Meanwhile, all upstream intermediate nodes that receive the RFN invalidate the particular route to avoid further packet losses. The sender remains in the snooze state until it is notified of the restoration of the route through a Route Reestablishment Notification (RRN) packet from an intermediate node. Then it resumes the transmission from the frozen state. The state machine of TCP-F is shown in Figure 5.6.

5.4.2.2 Explicit Link Failure Notification (ELFN)-Based Technique

The ELFN-based technique is similar to TCP-F. However in contrast to TCP-F, the evaluation of the proposal is based on a real interaction between TCP and the routing protocol. This interaction aims to inform the TCP agent about route failures when they occur. The authors use an ELFN message, which is piggybacked on the route failure message sent by the routing protocol to the sender. The ELFN message is like a "host unreachable" Internet Control Message Protocol (ICMP) message, which contains the sender's and receiver's addresses and ports, as well as the TCP packet's sequence number. On receiving the ELFN message, the source responds by disabling its retransmission timers and enters a "standby" mode. During the standby period, the TCP sender probes the network to check if the route is restored. If the acknowledgment of the probe packet is received, the TCP sender leaves the standby mode, resumes its retransmission timers, and continues the normal operations.

The probe interval of two seconds performs the best and makes this interval a function of the RTT instead of giving it a fixed value. For the RTO and Congestion

Window (CW) values upon route restoration, using the prior values before route failure performs better than initializing CW to one packet or RTO to six seconds.

5.4.2.2.1 Advantages

This technique provides significant enhancements over standard TCP, but further evaluations are still needed. For instance, different routing protocols should be considered other than the reactive protocol Dynamic Source Routing (DSR), especially proactive protocols such as Optimized Link State Routing (OLSR).

5.4.2.2.2 Disadvantages

In case of high load, ELFN performs worse than standard TCP, because ELFN is based on probing the network periodically to detect route reestablishment. Also, in the case of light loads, ELFN performs worse than standard TCP by 5 percent in the case of static ad hoc networks.

5.4.2.3 Ad Hoc TCP (ATCP)

ATCP utilizes network layer feedback, too. In addition to the route failures, ATCP tries to deal with the problem of high Bit Error Rate (BER). The TCP sender can be put into a persistence state, congestion control state, or retransmit state. The State Transition Diagram for ATCP at the sender is shown in Figure 5.7. The ATCP layer is inserted between the TCP and IP layers of the TCP source nodes. ATCP listens to the network state information provided by Explicit Congestion Notification (ECN) messages and by ICMP "Destination Unreachable" messages, then ATCP puts a TCP agent into the appropriate state. On receiving a Destination Unreachable message, a TCP agent enters a persist state. The TCP agent during this state is frozen, and no packets are sent until a new route is found by probing the network. The ECN is used as a mechanism to explicitly notify the sender about network congestion along the route being used. Upon the reception of ECN, TCP congestion control is invoked normally without waiting for a timeout event. To detect packet losses due to channel errors, ATCP monitors the received ACKs. When ATCP sees that three duplicate ACKs have been received, it does not forward the third duplicate ACK but puts TCP in the persist state and quickly retransmits the lost packet from TCP's buffer. After receiving the next ACK, ATCP will resume TCP to the normal state. Note that ATCP allows interoperability with TCP sources or destinations that do not implement ATCP. In the cases such as congestion, lossy links, partition, and packet reordering, the transfer time of a given file using ATCP yielded better performance than TCP. In addition to routes failure, ATCP tries to deal with the problem of high BER, network congestion, and packet reorder.

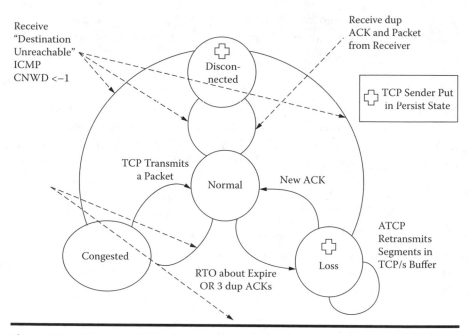

Figure 5.7 State transition diagram for ATCP at the sender.

5.4.2.3.1 Advantage

ATCP is a more robust proposal for TCP in MANETs.

5.4.2.3.2 Disadvantages

Some assumptions, such as ECN-capable node as well as sender node being always reachable, might be somehow hard to meet in a mobile ad hoc context. Also, the probing mechanism used to detect route reestablishment generates problems in the case of a high load.

5.4.2.4 TCP Buffering Capability and Sequencing Information (TCP-BuS)

TCP buffering capability and sequence information (TCP-BuS) uses the network feedback to detect route failure events and to take convenient reactions to this event. This introduces buffering capability in mobile nodes. This uses the source-initiated, on-demand, Associativity-Based Routing (ABR) protocol. The following enhancements are proposed:

Explicit notification: Two control messages are used to notify the source about the route failure and the route reestablishment. These messages are called Explicit

Route Disconnection Notification (ERDN) and Explicit Route Successful Notification (ERSN). On receiving the ERDN from the node that detected the route failure, called the pivoting node (PN), the source stops sending. And similarly, after route reestablishment by the PN using a localized query (LQ), the PN will transmit the ERSN to the source. On receiving the ERSN, the source resumes data transmission.

Extending timeout values: During the Route Reconstruction (RRC) phase, packets along the path from the source to the PN are buffered. To avoid timeout events during the RRC phase, the retransmission timer value for buffered packets is doubled.

Selective retransmission request: As the retransmission timer value is doubled, the lost packet along the path from the source to the PN is not retransmitted until the adjusted retransmission timer expires. To overcome this, an indication is made to the source so that it can retransmit this lost packet selectively.

Avoiding unnecessary requests for fast retransmission: When the route is restored, the destination notifies the source about the lost packets along the path from the PN to the destination. On receiving this notification, the source simply retransmits these lost packets. But the packets buffered along the path from the source to the PN may arrive at the destination earlier than the retransmitted packets. So, the destination will reply by duplicate ACK. These unnecessary request packets for fast retransmission are avoided.

5.4.2.4.1 Operation of TCP-BuS

Because ad hoc routes can be invalidated by node movements, we shall discuss the actions taken by TCP-BuS at the source, destination, and intermediate nodes to cope with host mobility.

5.4.2.4.2 TCP-BuS Functions at Source Node

At the source, TCP-BuS transmits its segments in the same manner as general TCP when there are no feedback messages (such as ERDN and ERSN messages). The slow start and congestion avoidance mechanisms function as usual. However, when the source receives the ERDN feedback message from the network, it stops sending data packets. In addition, it freezes all timer values and window sizes as in TCP-F. Next, the ERDN GEN SEQ value is extracted from the ERDN message, and the ERDN RCV SEQ is calculated. As an example (see Figure 5.8), a pivoting node detects a route failure when it has a segment (10) to transmit. The PN generates an ERDN message containing the sequence number (10) of the segment in the head of its transmit queue. Therefore, when the source receives the ERDN message, ERDN GEN SEQ is set to 10. Meanwhile, the source has been sending segments up to the segment (14).

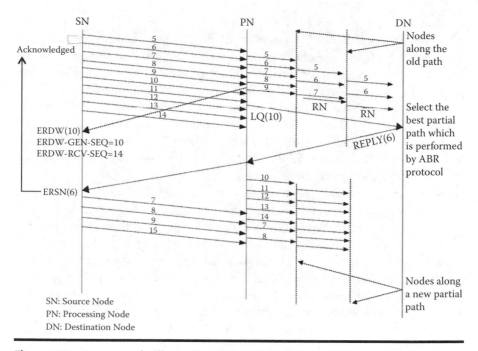

Figure 5.8 An example illustrating ERDN GEN SEQ and ERDN RCV SEQ mechanisms.

From this, we can calculate ERDN RCV SEQ, which is set to 14. Additionally, the next downstream node from PN will send the RN message toward the destination, which invalidates the old partial path and flushes buffered packets along that path. Because an ERDN message indicates that there is a route failure in the network, the source just waits for an ERSN message. On receiving this ERSN message, the source interprets that route reestablishment is successful. The source is then able to resume data transmission according to the TCP window-based mechanism. On receiving an ERSN message after a successful route reconstruction (i.e., after the PN node receives the REPLY message used in ABR, it generates the ERSN message with the last ACK parameter that is extracted from the REPLY message), the source can increase the congestion window by the amount of the acknowledged packets. At the same time, the source can safely assume that the segments whose sequence numbers range from (Last ACK + 1) to (ERDN GEN SEQ – 1) were discarded on the path from the PN to the destination in the network. It is obvious that these discarded packets should be retransmitted by the source. In Figure 5.8, because the ERSN message includes the segment sequence number (6) up to which the destination has received successfully, the source becomes aware that it should retransmit the segments from sequence number (7) to sequence number (9), which have been discarded along the old partial path. However, it depends on the congestion situation over the path from the source to the PN. An ERSN message can include congestion state information notifying the status of routers' queues at the intermediate nodes. One can make use of ICMP message such as Source Quench to indicate the presence of congestion, and the source will have to stop transmissions for a short period of time during which the intermediate node can catch up. Note that the timeout values at the TCP source for the unacknowledged and nonretransmitted segments from ERDN GEN SEQ to ERDN Receive Seqeunce (RCV SEQ) should be adjusted because of the expected increase in the packet arrival time at the destination due to the presence of route reestablishment. However, if packet losses are experienced on the partial path from the source to the PN due to congestion, the source reacts to the congestion and retransmits the lost packets selectively on receiving the selective retransmission request packet issued by the receiver. Therefore, it performs a congestion control procedure and reduces the congestion window size accordingly.

The parameters and control messages are as follows:

ERDN GEN SEQ: When an intermediate node detects a route failure and it cannot forward the buffered data packets, ERDN GEN SEQ is defined as the sequence number of the TCP segment pending in the head of line (HOL) of the node's transmit queue. ERDN GEN SEQ information is propagated from the PN to the source via the ERDN message.

ERDN RCV SEQ: When the source is transmitting TCP segments, if the source receives an ERDN message from the network, the source stops sending TCP segments. ERDN RCV SEQ is defined as the sequence number of the last TCP segment sent until the TCP source receives an ERDN control packet.

Last ACK: During route reconstruction, the destination responds to the LQ message with a REPLY. Last ACK is therefore defined as the sequence number of the last segment that the destination has received successfully.

By using ERDN GEN SEQ and ERDN RCV SEQ, mentioned above, the following information can be inferred:

■ The unacknowledged segments (buffered at the source) up to the segment whose sequence number is (ERDN GEN SEQ – 1) may be forwarded along the path from the node next to the PN toward the destination.

■ Unacknowledged segments (buffered at the source) whose sequence numbers are from ERDN GEN SEQ to ERDN RCV SEQ may be buffered at intermediate nodes.

5.4.2.4.3 TCP-BuS Functions at Intermediate Node

After a PN detects a route failure, it sends the ERDN message to notify the source of route failure and initiates partial route discovery using the LQ-REPLY process.

While the ERDN message is propagated toward the source, each intermediate node stops further transmission of data packets and buffers all pending packets to defer their transmission. After receiving the REPLY message, the PN notifies the source of successful route reestablishment via the ERSN message, which also includes the last ACK information. At each intermediate node receiving the ERSN message, transmission of buffered packets resumes.

5.4.2.4.4 TCP-BuS Functions at Destination Node

A receiver performs the normal TCP end-to-end procedure on the acquired path in case there is no route disconnection. Also, a selective retransmission mechanism as in TCP–selective ACK (SACK) can be applied for efficient flow control. We proposed an additional selective retransmission scheme to cope with lost packets due to congestion on the partial path from the source to the receiver. A request for selective retransmission of lost packets is generated at the receiver on detecting the absence of a consecutive segment sequence. This requires the source to react to the congestion. Given the above-mentioned approach, it is still possible that there are many requests for fast retransmission in the backward direction. Consider the case where segments having sequence numbers ranging from (Last ACK + 1) to (ERDN GEN SEQ – 1) will arrive at the destination later than those packets having sequence numbers ranging from ERDN GEN SEQ to ERDN RCV SEQ. As a result, the destination continues to request fast retransmission by sending duplicated ACK packets for each incoming packet received due to discrepancies in the sequence order. To avoid this problem, an additional procedure at the destination (see Figure 5.8) is required after receiving the LQ message. When receiving an LQ packet during a route recovery process, the destination extracts the ERDN GEN SEQ value piggybacked in the LQ packet. With the consideration of selective retransmission and avoiding the unnecessary requests for fast retransmission, the destination sends duplicated ACKs and requests for missing packets selectively according to the following rule. Here, we denote the sequence number of the incoming segment as "incoming SEQ." "Pivot value" is the sequence number whose subsequent segments are lost due to congestion. Therefore, the receiver notifies the source of lost segment information selectively. On receiving an LQ message for route extension, Pivot value = ERDN RCV SEQ. If incoming SEQ = ERDN GEN SEQ, then the transmission of duplicated ACKs for fast retransmission is refrained. If incoming SEQ > pivot value, notify the source of the information on missing segments. Pivot value = incoming SEQ. Otherwise, the transmission of duplicated ACKs is permitted.

TCP-BuS outperforms the standard TCP and the TCP-F under different conditions. TCP-BuS did not take into account that the pivoting node may fail to establish a new partial route to the destination. A comparison between different end-to-end approaches is shown in Table 5.1.

Table 5.1 Comparison between Different End-to-End Approaches

	TCP-F	ELFN	ATCP	TCP-BuS
High BER packet loss	Not handled	Not handled	Handled	Not handled
Route failures (RF) detection	RFN packet freezes TCP sender state.	ELFN packet freezes TCP sender state.	ICMP "destination unreachable" freezes TCP sender state.	ERDN packet freezes TCP sender state.
Route reconstruction (RR) detection	RRN packet resumes TCP to normal state	Probing mechanism	Probing mechanism	ERSN packet resumes TCP to normal state
Packet reordering	Not handled	Not handled	Handled	Not handled
Congestion window (CW) and retransmission timeout (RTO) after RR	Old CW and RTO	Old CW and RTO	Reset for each new route	Old CW and RTO
Reliable transmission of control messages	Not handled	Not handled	Not handled	Handled
Evaluation	Emulation; no routing protocol considered	Simulation	Experimental; no routing protocol considered	Simulation

5.4.3 Ad Hoc Transport Protocol (ATP)

The new transport protocol called ad hoc transport protocol (ATP) is tailored toward the characteristics of ad hoc networks. ATP, by design, is an antithesis of TCP and consists of: rate-based transmissions, quick-start during connection initiation and route switching, network-supported congestion detection and control, no retransmission timeouts, decoupled congestion control and reliability, and coarse-grained receiver feedback.

5.4.3.1 The ATP Design

This section outlines the key design elements of the proposed ATP.

5.4.3.1.1 Layer Coordination

One of the emerging trends in adapting protocols for wireless networks in general and ad hoc networks in particular is more involved coordination between different layers of the protocol stack. For example, most routing protocols designed for ad hoc networks rely on MAC layer information to detect link (and hence path) failures. A further degree of coordination possible in ad hoc networks is the explicit coordination between the different nodes in the network to improve end-to-end performance. For example, TCP-ELFN uses link failure notification from the intermediate routers to freeze TCP's state at the sender. One of the key cornerstones of the ATP design is the use of lower layer information and explicit feedback from other network nodes to assist in the transport-layer mechanisms. Specifically, ATP uses feedback from the network nodes for three different purposes:

1. Initial rate feedback for start-up rate estimation.
2. Progressive rate feedback for congestion detection, congestion avoidance, and congestion control.
3. Path failure notification. Although any node coordination can potentially constrain the scalability of a protocol, ATP does not require any per-flow state maintenance at the intermediate nodes and hence is highly scalable.

5.4.3.1.2 Rate-Based Transmissions

ATP uses rate-based transmissions in lieu of the window-based transmissions performed by TCP. Rate-based transmissions aid in improving performance in two ways:

- They avoid the drawbacks due to burstiness.
- Because the transmissions are scheduled by a timer at the sender, the need for self-clocking through the arrival of ACKs is eliminated.

The latter benefit is used by ATP to decouple the congestion control mechanism from the reliability mechanism and also to alleviate the impact of reverse-path characteristics on the performance experienced by the data stream on the forward path. Although an obvious limitation of rate-based schemes is the timer overheads incurred at the sender, the timer granularity required for the limited bandwidths in an ad hoc network is large enough to be realized without significant overheads. For example, with a reasonable load in the network, say, 10 flows (or 25 flows); a packet size of 512 bytes; and a raw channel capacity of two Mbps, the timer granularity required is 40 milliseconds (or 125 ms).

5.4.3.1.3 Decoupling of Congestion Control and Reliability

Unlike in TCP where the congestion control and reliability mechanisms are tightly coupled through dependence on ACK arrival, in ATP the two mechanisms are decoupled. Congestion control is performed using feedback from the network, and reliability is ensured through coarse-grained receiver feedback and selective ACKs.

- To facilitate congestion control, the intermediate nodes in the network provide congestion information in terms of the available rate. The feedback is piggy-backed on the data packets in the forward path, and the ATP receiver consolidates such information and sends back the collated feedback information.
- For reliability, the receiver also uses selective ACKs to report back to the sender any new holes observed in the data stream. Unlike in TCP where the SACK information is complementary to the cumulative ACK scheme, ATP relies solely on the SACK information.

5.4.3.1.4 Assisted Congestion Control

ATP's congestion control protocol relies on feedback from the intermediate nodes traversed by the connection to adapt the sending rate. Briefly, each node in the network maintains two parameters:

- Q_t (an exponential average of the queuing delay experienced by packets traversing that node)
- T_t (an exponential average of the transmission delay experienced by the HOL packet at that node)

T_t is influenced by the contention experienced between packets within nodes in the same contention vicinity, and Q_t is influenced by the contention between packets belonging to different flows at the same node. For every packet that passes through a node, the node stamps the sum $Q_t + T_t$ if the already stamped sum on the packet is smaller than its current value. The receiver of an ATP connection further performs an exponential averaging of the values stamped on the incoming packets. For every epoch period, the receiver sends rate feedback to the sender using the exponentially averaged value. The sender, based on its current rate and the rate specified in the feedback, determines whether to increase, decrease, or maintain its rate. The maintain phase in ATP is a critical difference from the states that a TCP connection can be in. In addition, the increase and decrease operations performed by ATP are more accurate because of the network feedback received.

5.4.3.1.5 TCP Friendliness and Fairness

TCP friendliness is not a constraint under which ATP is designed, because it is targeted for ad hoc network environments where network nodes will possess a dedicated protocol stack. However, fairness among ATP flows is still of key concern just as in TCP, because ATP relies on the intermediate network nodes for feedback on congestion.

5.4.4 The ATP Protocol

Just as in TCP, ATP primarily consists of mechanisms at the sender to achieve effective congestion control and reliability. However, unlike in TCP, ATP relies on feedback not just from the receiver, but also from the intermediate nodes in the connection path. In terms of specific functionality, the intermediate nodes provide congestion feedback to the sender, and the receiver provides feedback for both flow control and reliability. The receiver also acts as a collator of the congestion information provided by the intermediate nodes in the network before the information is sent back to the sender. The receiver provides the reliability, flow control, and collated congestion control information through periodic messages. The sender, on the other hand, is responsible for connection management, start-up rate estimation (with network feedback), congestion control, and reliability. In the rest of the section, we start with describing the role of the intermediate nodes. We then describe the mechanisms at the ATP receiver for providing rate feedback and sending SACK information to the sender. Finally, we describe in detail the different components of the ATP sender for start-up behavior, congestion control, and reliability.

5.4.4.1 Intermediate Node

ATP relies on the intermediate nodes that a connection traverses to provide rate feedback information. Intermediate nodes in the network maintain the sum of the average queuing delay (Q_t) and the average transmission delay (T_t) experienced by packets traversing through them. Whereas T_t at a node is impacted by the contention between the different nodes in the vicinity of that node, Q_t is impacted by the contention at that node between different flows traversing the node. Note that the values Q_t and T_t are computed over all the packets traversing the node, irrespective of the specific flow to which the packets belong. Thus, Q_t and T_t are maintained on a per-node basis, and not on a per-flow basis. For every outgoing packet, an intermediate node updates its Qt and T_t values. The values are maintained using exponential averaging as follows:

$$Q_t = \alpha + Q_t + (1 - \alpha) + Q_{sample} \qquad (5.1)$$

$$T_t = \alpha + T_t + (1 - \alpha) + T_{sample} \tag{5.2}$$

where Q_{sample} and T_{sample} are the queuing delay and transmission delay experienced by the outgoing packet. We use a value of 0.75 for α in our simulation results. In addition, each packet consists of a rate feedback field D (note that the rate will actually be an inverse of D) that consists of the maximum $Q_t + T_t$ value at the upstream nodes the packet has traversed through. When the packet is dequeued for transmission, the intermediate node checks to see if D is smaller than the $Q_t + T_t$ value at that node. If D is smaller, the intermediate node updates the D on the packet to its $Q_t + T_t$ value. When the receiver receives a packet, the D field in the packet indicates the maximum (average) delay experienced by packets at any of the intermediate nodes it traversed through.

When a connection starts up or a probe packet is sent by the sender, the intermediate node behavior is the same except when there is no other traffic around the node. When the node observes an idle channel, it uses $\eta + (Q_t + T_t)$ as the delay instead of the normal $Q_t + T_t$. The reasoning for this behavior is as follows. When the channel around the intermediate node is idle, the $Q_t + T_t$ values will be determined by the actual queuing delay and transmission delay experienced by the probe packet. However, when the actual data flow begins, packets belonging to the flow, at every hop in the path, will contend with other packets belonging to the same flow at both upstream and downstream nodes. Intermediate nodes project this rate by appropriately setting η. For Carrier Sense Multiple Access with Collision Avoidance (CSMA/CA), the typical value for η is 3. (In a linear chain, every third node can transmit, but for a large number of hops it can be as high as 5 for a path of length 5.) Note that ATP's requirement for node coordination thus constitutes only the maintenance of two parameters, Q_t and T_t, and appropriately stamping packets being forwarded. Also, the only additional field in the ATP header for the forward (data) path, besides the other fields in the TCP header, is the rate feedback field D.

5.4.4.2 ATP Receiver

The ATP receiver provides periodic feedback to the sender to assist in its reliability and flow control mechanisms. In addition, it also collates the rate feedback information provided by the intermediate nodes (through the D field on the packets), and sends it back to the sender. To send the feedback periodically, the receiver runs an epoch timer of period E. Note that period E should be longer than the round-trip time of a connection, but at the same time must be short enough to track the dynamics of the path characteristics. E is empirically chosen to be one second in our simulations.

5.4.4.2.1 Rate Feedback

For every incoming packet belonging to a flow, the receiver performs an exponential averaging of the D value specified in the packet:

$$D_{avg} = \beta + D_{avg} + (1 - \beta) + D \qquad (5.3)$$

After every epoch timer expiration, the receiver provides the D_{avg} value at that time as feedback to the sender. The exception is when the flow control rate determined by the receiver is smaller than the rate projected from the D_{avg} value. We use a value of 0.75 for β in all our simulations.

5.4.4.2.2 Reliability Feedback

The ATP receiver uses SACKs for providing information about losses in the data stream received. Because the feedback is not provided for every incoming data packet, but rather on a periodic basis, ATP uses a larger number of SACK blocks than TCP-SACK. Whereas TCP-SACK uses only 3 blocks per ACK, ATP uses 20 SACK blocks in its reliability feedback. For the typical channel data rate of 2 Mbps, the average per-flow rate is about 100 Kbps even when there are only 5 connections in the network. For an epoch period of one second and a packet size of 512 bytes, the rate translates into about 25 packets per second. Hence, when using twenty SACK blocks, most if not all of the losses in an epoch can be identified. However, unlike in TCP where the SACK blocks on ACKs progressively identify newer holes, in ATP the SACK blocks always identify the first sequence of 20 holes in the data stream. This is because ATP does not use a retransmission timeout at the sender and hence has to rely on the feedback from the receiver to perform correct error recovery.

In terms of the reverse-path overheads, for the above example, TCP-SACK can be shown to incur an overhead of 2.4 kbps for the SACK blocks (one ACK for every two data packets, three blocks per ACK, two sequence numbers per block, and four bytes per sequence number), whereas ATP would incur an overhead of 1.28 kbps for an epoch period of one second. However, the overhead of ATP can be shown to be considerably smaller if the overall header costs at the transport layer and at the lower layers are accounted for. This is because ATP aggregates its feedback into one packet, whereas TCP-SACK will send feedback for every two data packets.

5.4.4.2.3 Flow Control Feedback

Unlike in TCP, where the flow control is achieved through appropriate window advertisements, ATP performs flow control by observing the application read rate R_{app} at which the application is processing in sequence data from the receive buffer.

When rate feedback is being sent to the sender, if the R_{app} is smaller than the rate feedback, the rate feedback information is replaced with the R_{app}. The receiver, in its periodic feedback message to the sender (once every epoch), sends both the rate and reliability feedback.

5.4.4.3 ATP Sender

The ATP sender, like in TCP, consists of most of the driving mechanisms of the transport-layer protocol. Specifically, the ATP sender consists of the components for the following functionality:

- Quick-start
- Congestion control
- Reliability
- Connection management

5.4.4.3.1 Quick-Start

During connection initiation, or when recovering from a timeout, TCP's slow-start mechanism will take a few round-trip times before it can converge on the available bandwidth for a flow. Due to the frequent path failures and resultant timeouts in an ad hoc network, a TCP connection can end up spending a considerable portion of its lifetime in the slow-start phase, thus degrading network utilization. ATP uses a mechanism called "quick-start" to probe for the available network bandwidth within a single round-trip time. Essentially, during connection initiation, ATP uses a TCP-like SYN – SYN + ACK exchange between the sender and the receiver. The intermediate nodes, when they forward the SYN packet, stamp on the packet the $Q_t + T_t$ delay. When the receiver responds back with an ACK, it piggybacks onto the ACK the $Q_t + T_t$ value that was stamped on the incoming SYN packet. The sender, upon receiving the ACK, starts using the rate value obtained based on the feedback. ATP performs the quick-start operation both during connection initiation and when the underlying network path traversed by the connection changes. The motivation for performing quick-start when a path change occurs is straightforward. When a new path is used, the connection is not aware of the available bandwidth on the path. Hence, performing bandwidth estimation once again allows the connection to operate at the true available bandwidth instead of either overutilizing or underutilizing the resources available along the new path.

5.4.4.3.2 Congestion Control

Unlike TCP, which has a two-phase congestion control protocol with an increase phase and a decrease phase, ATP uses a three-phase congestion control protocol consisting of increase, decrease, and maintain phases. One of the key differences between TCP's congestion control mechanisms and those of ATP is the network feedback that ATP's mechanisms use. Because TCP does not rely on any network support, it probes for more bandwidth by linearly increasing the congestion window size at the sender. Similarly, when a loss occurs, because TCP does not know the true extent of congestion, it conservatively performs a multiplicative decrease of the congestion window size. ATP, on the other hand, relies on feedback from the intermediate network nodes. Hence, its increase can be more aggressive than that of TCP, its decrease can be less conservative than that of TCP, and more importantly it can operate in a maintain phase when network conditions do not change. The phases are as follows:

- *Increase phase*: When the feedback rate from the receiver is greater than the current rate S by a threshold φ + S, the sender enters the increase phase, where φ is a small constant used to prevent fluctuations. The threshold is kept as a function of the current rate to allow contending flows with lower rates to increase more aggressively than the flows with larger rates. Once an increase decision is taken, flows increase their rates only by a fraction k of the potential increase amount. We choose a value of 5 for k in the simulations due to the following rationale: when the rate of a flow is increased by one packet per second, the induced load (when the underlying MAC scheme is CSMA/CA) in the network can increase by up to five packets per second. For example, consider a path A–B–C–D–E–F. Even when the rate of a flow traversing this path increases by one packet per second, a transmission on the link C–D will contend with all the other four packet transmissions on the path.
- *Decrease phase*: On the other hand, when the feedback rate is smaller than the current rate, the sender performs the decrease phase merely by adjusting its current rate to the feedback rate.
- *Maintain phase*: If the available rate R lies within (S, S + φ + S), the sender maintains its rate. Thus, unlike in TCP which has to be in either the increase or decrease phases, ATP, given stable network conditions, can operate in a state of equilibrium.

Note that the above congestion control decisions can be taken by the ATP sender only when it receives the rate feedback from the receiver correctly. It is possible that the rate feedback from the receiver is lost due to path failures on the reverse path. ATP addresses this issue by performing a multiplicative decrease of the sending rate for every epoch in which it does not receive feedback from the receiver, up to a maximum of two epochs. If it does not receive any feedback at the end of the third epoch, the ATP sender goes into its connection initiation phase, sending one probe every epoch till it hears back from the receiver. Note that once it hears from the receiver, it will use the rate on the feedback packet for its transmissions as part of its quick-start mechanism.

5.4.4.3.3 Reliability

The receiver as part of its periodic feedback sends information about any holes in the data stream it has received. The ATP sender treats the SACK information just as in TCP by maintaining a SACK scoreboard data structure. Data marked to be retransmitted is sent with a higher preference than new data. Note that the congestion control mechanism in ATP is decoupled from the reliability mechanism. Hence, although the congestion control protocol determines the rate at which the sender should be sending, the reliability mechanism ensures that packets queued for retransmission are sent preferentially when the send timer expires. In other words, the retransmissions are performed within the regular transmission rate determined by the congestion control algorithm. In addition to the receiver informing the sender about losses, when there is a path failure, the ATP sender uses explicit link failure notification from the appropriate intermediate node. When such feedback is received, the ATP sender immediately enters the connection initiation phase as part of its recovery mechanism after a route switch. In the connection initiation phase, the ATP sender, for every epoch, sends a probe packet to the receiver. The probe packet is piggybacked on the next in-sequence data packet queued for transmission. Even if there are suffix losses due to the path failure, the subsequent periodic probe packets sent by the sender serve to elicit a feedback packet from the receiver containing the appropriate SACK information.

Note that the ATP sender will typically receive the link-failure notification before (if at all) it receives SACK information from the receiver. This is because the SACK information from the receiver will be generated only when there are no suffix losses because of the path failure. However, suffix losses will be the norm unless some route salvaging is done by the intermediate nodes. Even if both the link-failure notification and the receiver's periodic feedback packets are lost, the sender will eventually enter the probe phase due to lack of feedback from the receiver and hence recover from any losses.

5.5 Application-Controlled Transport Protocol (ACTP)

Unlike the TCP solutions discussed earlier in this chapter, ACTP is a lightweight transport-layer protocol. It is not an extension to TCP. ACTP assigns the responsibility of ensuring reliability to the application layer. It is more like UDP with feedback of delivery and state maintenance. ACTP stands in between TCP and UDP, where TCP experiences low performance with high reliability and UDP provides better performance with high packet loss in ad hoc wireless networks.

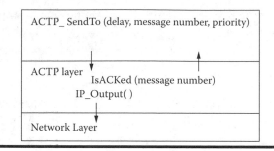

Figure 5.9   An illustration of interfaces functions used in ACTP.

The key design philosophy of ACTP is to leave the provisioning of reliability to the application layer and provide simple feedback information about the delivery status of packets to the application layer. ACTP supports the priority of packets to be delivered, but it is the responsibility of the lower layers to actually provide a differentiated service based on this priority. Figure 5.9 shows the ACTP layer and the applied programming interface (API) functions used by the application layer to interact with the ACTP layer. Each API function call to send a packet [*SendTo ( )*] contains the additional information required for ACTP such as the maximum delay the packet can tolerate (delay), the message number of the packet, and the priority of the packet. The message number is assigned by the application layer, and it need not be in sequence. The priority level is assigned for every packet by the application. It can be varied across packets in the same flow with increasing numbers referring to higher priority packets. The nonzero value in the message number field implicitly conveys that the application layer expects delivery status information about the packet to be sent. This delivery status is maintained at the ACTP layer and is available to the application layer for verification through another API function, "IsACKed<message number>." The delivery status returned by the "IsACKed<message number>" function call can reflect the following:

- A successful delivery of the packet (ACK received).
- A possible loss of the packet (no ACK received and the deadline has expired).
- Remaining time for the packet (no ACK received but the deadline has not expired).
- No state information exists at the ACTP layer regarding the message under consideration.

A zero in the delay field refers to the highest priority packet, which requires immediate transmission with minimum possible delay. Any other value in the delay field refers to the delay that the message can experience. On getting the information about the delivery status, the application layer can decide on retransmission of a packet with the same old priority or with an updated priority. Well after the packet's lifetime expires, ACTP clears the packet's state information and delivery status. The packet's lifetime is calculated as 4 × Retransmit TimeOut (RTO) and is set as the lifetime when the packet is sent to the network layer. A node estimates the RTO interval by using the round-trip time between the transmission time of a message and the time of reception of the corresponding ACK. Hence, the RTO value may not be available if there is no existing reliable connection to a destination. A packet without any message number (i.e., no delivery status required) is handled exactly the same way as in UDP without maintaining any state information.

5.5.1 Advantages and Disadvantages

One of the most important advantages of ACTP is that it provides the freedom of choosing the required reliability level to the application layer. Because ACTP is a lightweight transport-layer protocol, it is scalable for large networks. Throughput is not affected by path breaks as much as in TCP, as there is no congestion window for manipulation as part of the path break recovery. One disadvantage of ACTP is that it is not compatible with TCP. Use of ACTP in a very large ad hoc wireless network can lead to heavy congestion in the network as it does not have any congestion control mechanism.

5.6 Summary

This chapter discussed the major challenges that a transport-layer protocol faces in ad hoc wireless networks. The major design goals of a transport-layer protocol were listed, and a classification of existing transport-layer solutions was provided. TCP is the most widely used transport-layer protocol and is considered to be the backbone of today's Internet. It provides end-to-end, reliable, byte-streamed, in-order delivery of packets to nodes. Because TCP was designed to handle problems present in traditional wired networks, many of the issues that are present in dynamic topology networks such as ad hoc wireless networks are not addressed. This causes reduction of throughput when TCP is used in ad hoc wireless networks. It is very important to employ TCP in ad hoc wireless networks as it is important to seamlessly communicate with the Internet whenever and wherever it is available. This chapter provided a discussion on the major reasons for the degraded performance of traditional TCP in ad hoc wireless networks and explained a number of recently proposed solutions to improve TCP's performance.

5.7 Problems

5.1 Assume that when the current size of the congestion window is 48 kb, the TCP sender experiences a timeout. What will be the congestion window size if the next three transmission bursts are successful? Assume that the mobile satellite service (MSS) is one KB. Consider (1) TCP Tahoe and (2) TCP Reno.

5.2 Find out the probability of a path break for an eight-hop path, given that the probability of a link break is 0.2.

5.3 Discuss the effects of multiple breaks on a single path at the TCP-F sender.

5.4 What additional state information is to be maintained at the Forwarding Path (FP) in TCP-F?

5.5 Mention one advantage and one disadvantage of using probe packets for detection of a new path.

5.6 Mention one advantage and one disadvantage of using LQ and REPLY for finding partial paths in TCP-BuS.

5.7 What is the impact of the failure of proxy nodes in Split TCP?

5.8 During a research discussion, one of your colleagues suggests an extension of Split TCP where every intermediate node acts as a proxy node. What do you think would be the implications of such a protocol?

5.9 What are the pros and cons of assigning the responsibility of end-to-end reliability to the application layer?

5.10 What is the default value of β used for handling induced traffic in ATP, and why is such a value chosen?

Bibliography

V. Anantharaman and R. Sivakumar, A microscopic analysis of TCP performance analysis over wireless ad hoc networks (poster), in Proceedings of ACM SIGMETRICS 2002, Marina del Rey, CA, June 2002.

A. Bakre and B. Badrinath, I-TCP: Indirect TCP for mobile hosts, in Proceedings of 15th International Conference on Distributed Computing Systems (ICDCS), Vancouver, BC, Canada, May 1995.

B. Bakshi, P. Krishna, N. H. Vaidya, and D. K. Pradhan, Improving performance of TCP over wireless networks, in Proceedings of 17th International Conference on Distributed Computing Systems (ICDCS), Baltimore, MD, May 1997.

H. Balakrishnan, S. Seshan, E. Amir, and R. Katz, Improving TCP/IP performance over wireless networks, in Proceedings of ACM MOBICOM, Berkeley, CA, Nov. 1995.

K. Chandran, S. Raghunathan, S. Venkatesan, and R. Prakash, A feedback based scheme for improving TCP performance in ad-hoc wireless networks, in Proceedings of International Conference on Distributed Computing Systems, Amsterdam, May 1998, pp. 472–479.

T. Chen and M. Gerla, Global state routing: A new routing scheme for ad-hoc wireless networks, in Proceedings of IEEE ICC'98, Aug. 1998, pp. 171–175.

C. C. Chiang, H. K. Wu, W. Liu, and M. Gerla, Routing in clustered multihop mobile wireless networks with fading channel, in Proceedings of IEEE Singapore International Conference on Networks SICON'97, April 1997, pp. 197–212.

M. S. Corson and J. Macker, Mobile ad hoc networking (MANET): Routing protocol performance issues and evaluation considerations, Request for Comments 2501, IETF, Jan. 1999.

S. R. Das, R. Castaneda, and J. Yan, Comparative performance evaluation of routing protocols for mobile ad hoc networks.

T. D. Dyer and R. Bopanna, A comparison of TCP performance over three routing protocols for mobile adhoc networks, in Proceedings of ACM MOBIHOC 2001, Long Beach, CA, Oct. 2001.

L. M. Feeney, A taxonomy for routing protocols in mobile ad hoc networks, SICS Technical Report T99/07, Oct. 1999, http://citeseer.ist.psu.edu/feeney99 taxonomy.html.

M. Gerla, X. Hong, and G. Pei, Fisheye State Routing Protocol (FSR) for ad hoc networks, IETF Draft, 2001.

M. Gerla, K. Tang, and R. Bagrodia, TCP performance in wireless multi hop networks, in Proceedings of IEEE WMSCA, New Orleans, LA, Feb. 1999.

M. Handley, C. Bormann, B. Adamson, and J. Macker, NACK Oriented Reliable Multicast (NORM) Protocol building blocks, in Internet draft, RMT Working Group, draft-ietf-rmt-bb-norm-05.txt, March 2003.

T. Henderson and R. Katz, Satellite Transport Protocol (STP): An SSCOP-based transport protocol for datagram satellite networks, in Proceedings of 2nd Workshop on Satellite-Based Information Systems (WOSBIS), Budapest, Hungary, 1997.

G. Holland and N. H. Vaidya, Impact of routing and link layers on TCP performance in mobile ad-hoc networks, in Proceedings of IEEE WCNC, New Orleans, LA, Sept. 1999.

G. Holland and N. H. Vaidya, Analysis of TCP performance over mobile ad hoc networks, in Proceedings of ACM MOBICOM, Seattle, WA, Aug. 1999, pp. 219–230.

A. Iwata, C. C. Chiang, G. Pei, M. Gerla, and T. W. Chen, Scalable routing strategies for ad-hoc wireless networks, *IEEE Journal on Selected Areas in Communications*, 17(8):1369–1379, Aug. 1999.

P. Jaquet, P. Muhlethaler, and A. Qayyum, Optimized link state routing protocol, IETF Draft, 2001.

X. Jiang and T. Camp, A review of geocasting protocols for a mobile ad hoc network, Grace Hopper Celebration (GHC), 2002, http://toilers.mines.edu/papers/pdf/Geocast-Review.pdf.

C. E. Koksal and H. Balakrishnan, An analysis of short-term fairness in wireless media access protocols (poster), in Proceedings of ACM SIGMETRICS, Measurement and Modeling of Computer Systems, Santa Clara, CA, 2000, pp. 118–119.

J. Liu and S. Singh, ATCP: TCP for mobile ad hoc networks, *IEEE Journal on Selected Areas in Communications*, 19(7):1300–1315, July 2001.

E. L. Madruga and J. J. Garcia -Luna-Aceves, Scalable multicasting: The core-assisted mesh protocol, *Mobile Networks and Applications*, 6(2):151–165, March 2001.

J. P. Monks, P. Sinha, and V. Bharghavan, Limitations of TCP-ELFN for ad hoc networks, in Workshop on Mobile and Multimedia Communication, Marina del Rey, CA, Oct. 2000.

S. Murphy and J. J. Garcia-Luna-Aceves, An efficient routing protocol for wireless networks, ACM Mobile Networks and Applications, 183–197, Nov. 1996.

A. Nasipuri and S. Das, On-demand multipath routing for mobile ad hoc networks, in Proceedings of IEEE International Conference on Computer Communications and Networks (ICCCN), Boston, MA, Oct. 1999, pp. 64–70.

N. Nikaein, H. Labiod, and C. Bonnet, DDR distributed dynamic routing algorithm for mobile ad-hoc networks, in Proceedings of the First Annual Workshop on Mobile Ad Hoc Networks and Computing, MobiHOC 2000, Boston, Aug. 2000, pp. 19–27.

C. Parsa and J. J. Garcia-Luna-Aceves, Improving TCP performance over wireless networks at the link layer, *Mobile Networks and Applications*, 5(1):57–71, 2000.

C. E. Perkins and E. M. Royer, Ad-hoc on-demand distance vector (AODV) routing, in MANET Working Group. IETF, Internet draft, draft-ietf-manet-aodv-12.txt, Nov. 2002.

C. E. Perkins, ed., *Ad hoc networking*, Addison-Wesley, Reading, MA, 2001.

A. Qayyum, L. Viennot, and A. Laouiti, Multipoint relaying: An efficient technique for flooding in mobile wireless networks, INRIA research report RR-3898, 2000, http://citeseer.ist.psu.edu/qayyum00multipoint.html.

E. M. Royer and C-K. Toh, A review of current routing protocols for ad-hoc mobile wireless networks, *IEEE Personal Communications*, 46–55, April 1999.

P. Sinha, N. Venkitaraman, R. Sivakumar, and V. Bharghavan, WTCP: A reliable transport protocol for wireless wide-area networks, in Proceedings of ACM MOBICOM, Seattle, WA, Aug. 1999.

R. Sivakumar, P. Sinha, and V. Bharghavan, CEDAR: A core-extraction distributed ad hoc routing algorithm, *IEEE Journal on Selected Areas in Communications*, 17(8), Aug. 1999.

M. W. Subbarao, Mobile ad hoc data networks for emergency preparedness telecommunications: Dynamic power-conscious routing concepts, submitted as an interim project for Contract Number DNCR086200 to the National Communications Systems, 2000.

M. W. Subbarao, Ad hoc networking critical features and performance metrics, white paper, National Institute of Standards and Technology, Sept. 1999.

M. W. Subbarao, Dynamic power-conscious routing for MANETs: An initial approach, in Proceedings of IEEE VTC Fall 1999, Amsterdam, 1999.

C-K. Toh, *Ad hoc mobile wireless networks: Protocols and systems*, Prentice Hall PTR, Upper Saddle River, NJ, 2002.

C-K. Toh, Associativity-based routing protocol to support ad-hoc mobile computing, *Wireless Personal Communications*, special issue on mobile networking and computing systems, 4(2)103–139, March 1997.

A. Vahdat and B. Becker, Epidemic routing for partially-connected ad hoc networks, Duke technical report CS-2000-06, July 2000, Duke University, Durham, NC.

Chapter 6

Quality of Service (QoS) in Ad Hoc Networks

6.1 Introduction to QoS

Quality of service (QoS) is a measure of the level of service that a particular data gets in the network. The network is expected to guarantee a set of measurable pre-specified service attributes to the users in terms of end-to-end performance such as delay, bandwidth, probability of packet loss, delay variance (jitter), and so forth. Power consumption is another QoS attribute which is more specific to mobile ad hoc networks (MANETs).

Traditional Internet QoS protocols like Resource Reservation Protocol (RSVP) cannot be easily migrated to the wireless environment due to the error-prone nature of wireless links and the high mobility of mobile devices. This is true for mobile ad hoc networks where every node moves arbitrarily, causing the multihop network topology to change randomly and at unpredictable times.

6.1.1 QoS in Different Layers

- *Application layer*: Requests a particular Class of Service (CoS) and adaptively changes its CoS depending on the network state (using feedback from the network) so that its packets are not rejected if the network fails to keep up with the desired CoS.

- *Session or presentation layers*: These can act as classifiers to distinguish between the different CoS and map data belonging to different classes to separate queues for the transport layer.
- *Transport layer*: Services the higher CoS queue more often using some arbitration, and maps the three classes of service to different network-layer channels (a different routing protocol for each CoS).
- *Network layer*: Seeing each CoS data differently, it uses different routing mechanisms for each of them and sends the data into three different queues for the data-link layer.
- *Data-link layer*: Uses different protocols again for different CoS so that faster, better, and assured service can be delivered to, for example, the Gold CoS. It will also have to take care of the Medium Access Control (MAC) issues to make sure that the CoS can be delivered.
- *Physical layer*: This again may use better error-correction schemes and may give more time for transmission to higher CoS. Also, it may adaptively try to change the modulation scheme and so forth, with respect to the state of the channel, and hence try to provide service even when the channel Bit Error Rate (BER) is high.

6.1.2 QoS Analysis

In the previous section, we looked at each of the layers and their contribution toward ensuring QoS guarantees in the network. We can also look at the issue from the QoS provider's point of view.

6.1.2.1 QoS Model

The QoS model specifies the architecture in which certain services could be provided in the network. A QoS model for MANETs should first consider the challenges of MANETs, for example dynamic topology and time-varying link capacity.

6.1.2.2 QoS Resource Reservation

Each flow with some QoS guarantee will have to be allocated some resources which will be exclusively for its use. This will ensure that as soon as the packet of that particular flow comes, it will not have to wait for some path or resource to be freed and it will be transmitted to the next node "instantly." Another issue that comes up here is what resources should be reserved for a given set of QoS parameters and how they will be freed when the flow is terminated. Signaling that a particular set of resources has been allocated is also a major issue.

6.1.2.3 QoS Routing

QoS routing needs to be ensured by the network layer. The idea is to find a route to the destination which satisfies the QoS requirements. Such a routing is sometimes termed as "QoS-aware routing." This forms a very important because nodes are mobile, connections will break and will be made dynamically, and, hence, the routing protocol needs to take care of such situations.

6.1.2.4 QoS Medium Access Control Protocol

Some medium access guarantee should be provided by the data-link layer for supporting the given QoS requirements. This plays a very important role in fulfilling the desired objectives because the transmitting node should get the required transmission time so that it can send the data to the next node. Collision Avoidance (CA), hidden terminal, and exposed terminal problems need to be solved.

6.2 Issues and Challenges Involved in Providing QoS

QoS can be measured in terms of parameters like data rate, delay, delay variation (jitter), and packet loss. Providing QoS in MANETs has its own challenges and problems.

6.2.1 Challenges to Be Faced

- *Varying physical link properties*: Because the wireless link is unpredictable and time varying, it becomes difficult to ensure that a minimum level of service is satisfied.
- *Medium access issues*: Because the wireless channel is shared by many devices, managing them in such a way that the QoS guarantees are fulfilled is difficult.
- *Routing*: Because the nodes are mobile, the network topology changes randomly with time, and the routing protocol needs to update the routes and links.
- *Power consumption*: Nodes, being mobile, have a limited power capacity.
- *Characterization of the link state*: Because the network state changes with time, one needs to have some mechanism in place which can continue to update the network state and, based on it, predict whether it will be able to achieve a particular quality of service requirement or not.
- *Dynamic topologies*: Nodes are free to move arbitrarily; thus, the network topology which is typically multihop may change randomly and rapidly at unpredictable times, and may consist of both bidirectional and unidirectional links.

6.2.2 Issues and Design Considerations

6.2.2.1 Adaptive Services for Continuous Media Flow

The most suitable service paradigm for mobile ad hoc networks is adaptive in nature. We observe that adaptive voice and video applications operating in mobile cellular networks are capable of responding to packet loss, delay jitter, changes in unavailable bandwidth, and handoff while maintaining some level of service quality. Although adaptive multimedia applications can respond to network dynamics, they typically require some minimum bandwidth assurance below which they are rendered useless. In this context, adaptive services provide minimum bandwidth assurances to real-time voice and video flows and data, allowing for enhanced levels (i.e., maximum bandwidth) of service to be delivered when resources become available.

6.2.2.2 Separation of Routing, Signaling, and Forwarding

Here the routing protocols interwork with resource management to establish paths through the network that meet end-to-end QoS requirements. In this case, there is a certain level of integration of resource management and routing. One could apply such an approach to MANET routing protocols given that the time scales over which new routes are computed are much faster than those traditionally found in the case of routing in fixed infrastructures. Session setup and routing (i.e., computing new routes) are distinct and functionally independent tasks. Therefore, we believe that signaling, resource management, and routing should be modeled independently in the network architecture.

6.2.2.3 In-Band Signaling

In-band signaling systems are capable of operating close to packet transmission speeds and are therefore well suited toward responding to fast-time scale dynamics found in mobile ad hoc environments, as illustrated in Figure 6.1. The term "in-band signaling" means carrying the control information along with data, whereas the term "out-of-band signaling" means carrying the control information in separate control packets and channels that are different from the data path.

6.2.2.4 Soft-State Management

The "soft-state" approach involves state management at intermediate routing nodes which is suitable for the management of reservations in mobile ad hoc networks. Soft state models the transient nature of network reservation, which has to be responsive

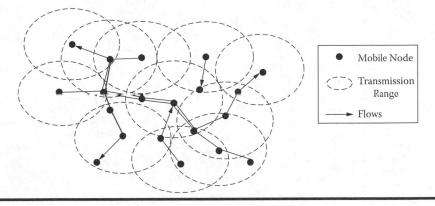

Figure 6.1 Mobile ad hoc networking.

to fast time-scale wireless dynamics, moderate time-scale mobility changes, and longer time-scale session "holding times."

Hence, the development of new QoS frameworks and protocols based on the notion of in-band signaling and soft-state management, and constructed with the separation of routing, QoS, signaling, and forwarding functions, will provide a responsive, scalable, and flexible solution to delivering adaptive services in mobile ad hoc networks.

6.3 Classification of QoS Solutions

QoS solutions are basically classified into QoS approach and the layers at which they operate. QoS approaches can be classified into the following types.

Coupled: Here, there will be close interaction between the routing algorithm and the QoS mechanism for providing QoS guarantees.

Examples: Ticket-Based Probing (TBP), Predictive Location-Based QoS Routing Protocol (PLBQR), Time Domain Reflectometry (TDR), Quality of Service Ad Hoc On-Demand Distance Vector (QoS-AODV), BR, OQR, OLMOR, Active Query Router (AQR), Core-Extraction Distributed Ad Hoc Routing (CEDAR), and Intelligent Optimization Self-Regulated Adjustment (INORA)

Decoupled: Here, there will not be any routing protocol on which the QoS mechanism is dependent.

Examples: INSIGNIA, stateless wireless ad hoc network (SWAN), and PRTMAC

Independent: Here, the network-layer protocols are not dependent on MAC.

Examples: TBP, PLBQR, QoS-AODV, INSIGNIA, INORA, and SWAN

Dependent: The network-layer protocols are dependent on the MAC layer.

Examples: TDR, BR, OQR, OLMQR, AQR, CEDAR, and PRTMAC

Table driven: There will be the routing table in each node which helps in transmitting the packets.

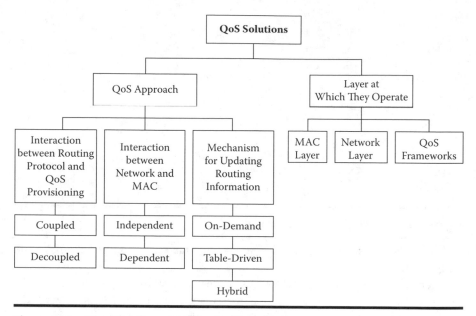

Figure 6.2   Classification of QoS solutions.

Example: PLBQR

On demand: The source node will find the route. No routing tables are provided.

Examples: TBP, TDR, QoS-AODV, OQR, OLMQR, AQR, INORA, and PRTMAC

Hybrid: This includes the features of both the table-driven and on-demand approaches.

Examples: BR and CEDAR

The classification of QoS solutions is depicted in Figure 6.2.

6.3.1   Medium Access Control (MAC)-Layer QoS Solutions

QoS-supporting components at upper layers, such as QoS signaling and QoS routing, assume the existence of a MAC protocol, which solves the problems of medium contention, supports reliable unicast communication, and provides resource reservation for real-time traffic in a distributed wireless environment. A lot of MAC protocols have been proposed for wireless networks. Unfortunately, their design goals are usually to solve medium contention and hidden and exposed terminal problems and to improve throughput. Most of them do not provide resource reservation and QoS guarantees to real-time traffic. The first problem that a MAC protocol in wireless networks should solve is the hidden or exposed terminal problem. Second, a QoS MAC protocol must provide resource reservation and QoS guarantees to real-time traffic. Following are the MAC protocols that guarantee QoS.

6.3.1.1 Multiple Access Collision Avoidance with Piggyback Reservation (MACA/PR)

MACA/PR is used as a MAC protocol for multihop wireless networks. MACA/PR provides rapid and reliable transmission of nonreal-time datagrams as well as guaranteed bandwidth support to real-time traffic.

For the transmission of non-real-time datagrams in MACA/PR, a host with a packet to send must first wait for a free "window" in the reservation table (RT), which records all reserved send and receive windows of any station within the transmission range. It then waits for an additional random time on the order of a single-hop round-trip delay. If it senses that the channel is free, it proceeds with a Request-To-Send (RTS)–Clear To Send (CTS)–Data Packet (PKT)–Acknowledgment (ACK) packet dialogue for a successful packet transmission. If the channel is busy, it waits until the channel becomes idle and repeats the above procedure.

For the transmission of real-time packets, the behavior of MACA/PR is different. To transmit the first data packet of a real-time connection, the sender S initiates an RTS–CTS dialogue and then proceeds with PKT–ACK dialogues if the CTS is received. For subsequent data packets (not the first one) of a real-time connection, only PKT–ACK dialogues are needed. Note that if the sender fails to receive several ACKs, it restarts the connection with the RTS–CTS dialogue again. MACA/PR does not retransmit the real-time packets after collision.

To reserve bandwidth for real-time traffic, the real-time scheduling information is carried in the headers of PKTs and ACKs. The sender S piggybacks the reservation information for its next data packet transmission on the current PKT. The intended receiver D inserts the reservation in its reservation table (RT) and confirms it with the ACK to the sender. The neighbors of the receiver D will defer their transmission once receiving the ACK. In addition, from the ACK, they also know the next scheduled receiving time of D and avoid transmission at the time when D is scheduled to receive the next data packet from S. The real-time packets are protected from hidden hosts by the propagation and maintenance of Reservation Tables (RT) among neighbors, not by the RTS–CTS dialogues. Thus, through the piggybacked reservation information and the maintenance of the reservation tables, the bandwidth is reserved and guaranteed for the real-time traffic.

6.3.1.2 RTMAC

RTMAC is an extension of the Destination-Sequenced Distance Vector (DSDV). It is responsible for finding an end-to-end path that satisfies the QoS bandwidth requirements. Bandwidth reservation for Constant Bit Rate (CBR) traffic is provided by dividing the transmission time into successive superframes. This scheme can also be extended to support Variable Bit Rate (VBR) traffic. The main concept in this approach is the flexibility of slot placement in the superframe. Each superframe consists of a sequence of reservation slots (resv-slots). The time duration of a resv-slot is twice the maximum propagation delay. Each session between source and destination nodes requires the reservation of a block of consecutive resv-slots. A node must first reserve a set of resv-slots and a guard band to cushion the propagation delay (henceforth, it is referred to as a "connection slot") on a superframe, and uses the same connection slot to transmit in successive superframes. A reservation table must be maintained by each node. The table contains information such as sender, receiver, and starting and ending times of the reservations that are active within its transmission range. This scheme is different from that of the time division multiple access (TDMA) environment because it requires no time synchronization (no need to maintain a global clock with the associated communication overhead) and uses a relative time for all reservation purposes. Each node transmits its reservation table along with the route update packet of DSDV.

The protocol includes the capability of a node to designate a specific connection slot to be reserved for a particular connection, which gives the routing protocol the flexibility to position the connection slot. The protocol applies different schemes to the reserve connection slot. Different schemes can be used to allocate connection slots such as first fit (reserve slot in the immediate freely available connection slot), best fit (place connection slots at a place that succeeds the connection slot on which the node receives the real-time packets), and fair fit (reserve connection slots in a way which creates free slots that can be used for best-effort traffic). A source node desiring to transmit data to a certain destination node checks the reservation information of its neighbors, finds free slots that can be reserved, and initiates the reservation process for those free slots.

6.3.1.3 Distributed Bandwidth Allocation/Sharing/ Extension (DBASE) Protocol

The DBASE Protocol is proposed. It supports multimedia traffic with CBR and VBR traffic over ad hoc wireless local area networks (WLANs). This protocol functions at the MAC layer. It is asynchronous and uses RTS–CTS–asynchronous DATA–ACK exchanges for channel access. In DBASE, real-time (rt) stations can reserve and free channel resources dynamically. The bandwidth allocation procedure is based on a contention (and backoff) process that only occurs before the first successful access and a reservation process after the successful contention. When the rt station leaves, the bandwidth is immediately released by DBASE.

6.3.2 Network-Layer QoS Solutions

QoS routing protocols search for routes with sufficient resources for QoS requirements. The QoS routing protocols should work together with resource management to establish paths through the network that meet end-to-end QoS requirements, such as delay or delay jitter bounds, bandwidth demand, or multimetric constraints. The QoS metrics could be concave or addictive.

QoS routing is difficult in MANETs for the following reasons:

- The overhead of QoS routing is too high for the bandwidth-limited MANETs because the mobile host should have some mechanisms to store and update link state information. We have to balance the benefit of QoS routing against the bandwidth consumption in MANETs.
- Because of the dynamic nature of MANETs, maintaining the precise link state information is very difficult.
- The traditional meaning that the required QoS should be ensured once a feasible path is established is no longer true. The reserved resource may not be guaranteed because of the mobility-caused path breakage or power depletion of the mobile hosts. QoS routing should rapidly find a feasible new route to recover the service.

There are three different kinds of network-layer solutions: (1) on-demand, (2) table-driven, and (3) hybrid.

6.3.2.1 Ticket-Based Probing (TBP)

The basic idea here is using tickets to limit the number of candidate paths. When a source wants to find QoS paths to a destination, it issues probe messages with some tickets. The number of the tickets is based on the available state information. One ticket corresponds to one path searching, and one probe message should carry at least one ticket. So the maximum number of the searched paths is bounded by the tickets issued from the source. When an intermediate host receives a probe message with n tickets, based on its local state information, it decides whether and how to split the n tickets and where to forward the probe(s). When the destination host receives a probe message, a possible path from the source to the destination is found.

6.3.2.2 QoS-Ad Hoc On-Demand Distance Vector (AODV)

This protocol is an extension of the well-known AODV protocol. It is on demand and designed to work in a TDMA network. This protocol combines information from both the network and data-link layer. Unlike other protocols which make path bandwidth calculations only after paths to the destination have been discovered, QoS-AODV incorporates path finding with the bandwidth reservation mechanism. QoS-AODV is fully aware of the bandwidth resource availability by coupling together routing and MAC TDMA layers. The nodes compete for the slots contained in the data phase of the TDMA frame. For the source node to send data to a destination node, it must establish a virtual circuit (VC) connection with that destination. The VC establishment process includes route discovery, path bandwidth calculation, and bandwidth reservation (data-phase-slot reservation) components. Each node keeps a schedule that contains information about both its own and its neighbor's time slots that are used for sending and receiving. A "schedule" is defined as a sequence of 1's and 0's where a number is the order of the corresponding slot in the data phase of the TDMA frame. The algorithm is used by each node to determine which slots are available to send to and receive from its neighbor, and to calculate link bandwidth scheduling from itself to each of its neighbors. The link bandwidth information is used in the calculation of the path bandwidth schedules to source and destination nodes. Modified AODV Hello messages are used which include slot-scheduling information. The Hello messages are sent either periodically or when link bandwidth information is changed.

In QoS-AODV, path discovery is done in the following manner. A source node that wants to send data to a particular destination determines if it has enough link bandwidth available to any of its neighbors. If it does not, it then denies the request initiated by its application layer. Otherwise, it creates a routing table entry for the requested application call ID and the destination address. Note that, in QoS-AODV, there is an entry in the routing table for each application call ID–source–destination triple instead of one per source–destination duplex as in the original AODV. The source node then sends the reservation request message (RREQ), which contains the call ID and the number of slots required for reservation, in addition to the standard AODV information.

When an intermediate node receives an RREQ message, it checks whether it already has an entry in its routing table corresponding to the received application call ID. The node then calculates the path bandwidth schedules using algorithms. If the calculated path bandwidth to the source is insufficient, then the node does not forward the RREQ message. Otherwise, the intermediate node augments the RREQ message with path and link bandwidth parameters, and broadcasts it further. The link bandwidth between two nodes is calculated as the intersection of their free slot schedules. The send link bandwidth (say, of a link AB at node A) is defined as the intersection of the free send slot schedule of the sender node (A) and the free receive slot schedule of the receiver node (B). The receive link bandwidth (say, of link AB at node A) is defined as the intersection of the free receive slot schedule of the receiver node (A) and the free send slot schedule of the sender node (B). In addition to information corresponding to the original AODV protocol, the route table entry contains the addresses of three.

Figure 6.3 represents a QoS-AODV propagation of the RREQ message from source node A to destination node E. The RREQ message is forwarded only if the required bandwidth is available at each link. At each intermediate node, it is augmented with QoS bandwidth information nodes along the path to the source, and link and path bandwidth schedules between those nodes. This information is needed to allocate slots that do not cause interference.

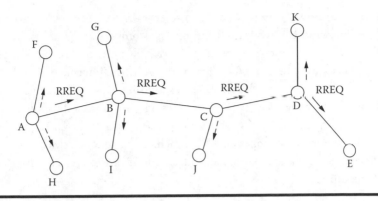

Figure 6.3   QoS-AODV.

An example of QoS-AODV route discovery is shown in Figure 6.3. Source node A needs to create a new VC to send data to destination node E. Node A broadcasts an RREQ message that contains the call ID and the number of slots required for the QoS path. Upon receiving the RREQ message, node B, knowing that node A is one of its neighbors, determines that the available path bandwidth from A to B is equal to the receive link bandwidth from A to B. Path bandwidth AB is calculated as a portion of the receive link bandwidth AB. Node B then augments the RREQ message with the calculated link bandwidth AB and the address of A, and rebroadcasts the RREQ message to all of its neighbors. When node C receives the propagated RREQ message from B, it knows that A is B's neighbor. Therefore, it calculates the available path bandwidth using the AB and BC receive link bandwidth to avoid any interference conflicts, including the hidden terminal problem. If the path bandwidths AB and BC contain the required number of slots, then C augments the RREQ message with the address of A and B, the receive link bandwidth BC, and the path bandwidth AB before it rebroadcasts it. Node D then receives the RREQ message; it calculates the path bandwidth to A using the link bandwidths AB, BC, and CD. If the calculated path bandwidth is sufficient, then D rebroadcasts the RREQ message after adding the address of C and B along with the receive link bandwidth CD and path bandwidth BC. When the destination node E receives the message, it uses the same algorithms to determine the path bandwidth scheduling CD and DE. Once the destination node E determines that there is enough bandwidth to the source node, it starts the reservation process by creating a reservation instance. The reservation parameters stored at each node along the VC for that VC ID include the following:

1. Source and destination IDs
2. Application call ID
3. Next-hop, previous-hop, and next-hop with bandwidth scheduling
4. Reservation status

The destination node (E in this example) reserves MAC receive slots corresponding to previous-hop bandwidth scheduling and composes a Reservation Message (RSV), which is a new message added to the AODV protocol. The RSV message is propagated back to the source, A in this example, by the intermediate nodes (B, C, and D in this example), which reserve corresponding MAC receive and send slots. When the source node receives the RSV message, it informs its application layer of the establishment of a VC connection to the destination. The algorithm also defines an Unreserved Message (URSV), which can be used to release slot resources if multiple reservations at a particular node are done due to race conditions caused by node mobility.

6.3.2.3 Core-Extraction Distributed Ad Hoc Routing (CEDAR)

The CEDAR algorithm can react effectively to the dynamics in MANETs. It includes three main components: core extraction, link state propagation, and route computation.

6.3.2.3.1 Core Extraction

The dominating set (DS) of a network is a set of hosts, such that every host in the network is either in the DS or a neighbor of a node in the DS. The minimum set of the DS is called the "minimum dominating set of the network."

The purpose of core extraction is to elect a set of hosts to form a core of the network by using only local computation and local state. The core of the network is an approximation of a minimum DS of the network. Every host in the DS is called a "core host." Every host not in the DS chooses one of its neighbors that are in the DS as its dominator. Note that the dominator of a core host is itself. Two core hosts are called "nearby core hosts" if the distance between them is no more than three. A path between two core hosts is called a "virtual link." The graph that consists of core nodes and virtual links to connect nearby core hosts is called a "core graph." A core path is a path in the core graph.

CEDAR presents a distributed algorithm to choose core nodes. When a host loses connectivity with its dominator due to mobility, it finds a core neighbor as its dominator, nominates one of its non-core-host neighbors to join the core, or itself joins the core. Because flooding in MANETs causes repeated local broadcasts, it is highly unreliable because of the abundance of hidden and exposed hosts. CEDAR proposes the core broadcast mechanism to ensure that each core host does not transmit a broadcast packet to every nearby core host. The core broadcast approach has very low overhead and adapts easily to topology changes. It also provides an efficient way to update link state information.

6.3.2.3.2 Link State Propagation

To compute the feasible QoS paths in CEDAR, each core host maintains its local topology as well as the link state corresponding to stable high-bandwidth links further away. Note that it does not keep the link state information of unstable or low-bandwidth further links, because these links are not useful in searching for the QoS routes. To achieve this goal, CEDAR utilizes the increase or decrease waves.

For every link l = (a, b), hosts a and b are responsible for monitoring the available bandwidth on the link. When the link l comes up or the bandwidth of the link l increases beyond a given threshold value, hosts a and b will notify their dominators to initialize a core broadcast for an increase wave, which indicates the stable high-bandwidth link. On the other hand, if the link l breaks down or the bandwidth of the link l decreases beyond a given threshold value, hosts a and b inform their dominators to initialize a core broadcast for a decrease wave, which indicates the unstable or low-bandwidth link. The increase wave is slow moving, whereas the decrease wave is fast moving. For the same link state, the fast-moving decrease wave will take over and kill the slow-moving increase wave. Finally, the survivable increase wave will propagate the stable high-bandwidth link state information through the cores. In addition, CEDAR provides a mechanism that keeps the decrease wave from propagating throughout the whole network. So the unstable low-bandwidth link states are kept locally.

6.3.2.3.3 Route Computation

The QoS route computation in CEDAR includes three main steps:

1. Discovering the location of the destination and establishing a core path to the destination
2. Searching for a stable QoS route with the established core path as a directional guideline
3. Dynamically recomputing QoS routes upon link failures or topology changes

When a source s wants to send messages to a destination d, it first sends an s–d–b triple to its dominator, dom(s), where b is the required bandwidth. If dom(s) can calculate a feasible path to d with its local state information, it responds to s immediately. Otherwise, dom(s) discovers the dom(d) if it does not know the location of d, and simultaneously establishes a core path to d. A core path request message is initialized and core-broadcasted by dom(s). By the virtue of the core broadcast algorithm, the core path request message traverses an implicitly established source-routed tree from dom(s), which is typically a breadth-first search tree. Thus the core path is approximately the shortest path in the core graph from dom(s) to dom(d), and provides a good directional guideline for the calculation of QoS routes.

Because dom(s) knows the up-to-date local topology and only some possibly out-of-date link state information about remote stable high-bandwidth links, dom(s) may not be able to calculate a possible path to d with enough required bandwidth based on its own link state knowledge. However, as mentioned before, the core path from dom(s) to dom(d) provides a good directional guideline for the possible QoS routes. Based on its own link state information, dom(s) will try to calculate a route with enough bandwidth to meet the bandwidth requirement to the furthest core node, dom(t), which is on the core path from dom(s) to dom(d). Then dom(s) sends dom(t) a message to notify it of continuing the same computation further. If dom(t) can calculate a route with enough bandwidth to d based on its own link state knowledge, then the computation is finished and a feasible path with enough bandwidth from dom(s) to d is found. Otherwise, dom(t) repeats the same operation as dom(s). The computation will continue along the core path from dom(s) to dom(d) step by step. Finally, at a core node tn, a feasible path with enough bandwidth from tn to d is found, or no possible path could be produced at tn. In the first case, the whole feasible path is the concatenation of the partial paths computed by the core nodes s, t, and tn. In the latter case, the bandwidth requirement cannot be satisfied and the request is rejected.

CEDAR deals with link failures by two mechanisms: (1) dynamic recomputation of a feasible route at the point of failure, and (2) notification back to the source to activate recomputation of a feasible route at the source. The two mechanisms work in concert to respond to topology changes.

6.3.3 QoS Model

The QoS model specifies the architecture in which certain services could be provided in the network. A QoS model for MANETs should first consider the challenges of MANETs, for example dynamic topology and time-varying link capacity. In addition, the potential commercial applications of MANETs require a seamless connection to the Internet. Thus, the QoS model for MANETs should also consider the existing QoS architectures in the Internet. In this section, we first introduce the QoS models for the Internet, such as IntServ and DiffServ. Then, a newly proposed QoS model is described for MANETs.

6.3.3.1 Integrated Service (IntServ) and Resource Reservation Protocol (RSVP) on Wired Networks

The basic idea of the IntServ model is that the flow-specific states are kept in every IntServ-enabled router. A flow is an application session between a pair of end users. A flow-specific state should include bandwidth requirement, delay bound, and cost of the flow. In addition to its Best Effort Service, IntServ proposes two service classes, Guaranteed Service and Controlled Load Service. The Guaranteed Service is provided for applications requiring fixed-delay bound. The Controlled Load Service is for applications requiring reliable and enhanced best-effort service. Because every router keeps the flow state information, the quantitative QoS provided by IntServ is for every individual flow. In an IntServ-enabled router, IntServ is implemented with four main components: the signaling protocol, the admission control routine, the classifier, and the packet scheduler. Other components, such as the routing agent and management agent, are the original mechanisms of the routers and can be kept unchanged. The RSVP is used as the signaling protocol to reserve resources in IntServ. Applications with Guaranteed Service or Controlled Load Service requirements use RSVP to reserve resources before transmission. Admission control is used to decide whether to accept the resource requirement. It is invoked at each router to make a local accept or reject decision at the time that a host requests a real-time service along some paths through the Internet. Admission control notifies the application through RSVP if the QoS requirement can be granted or not. The application can transmit its data packets only after the QoS requirement is accepted. When a router receives a data packet, the classifier will perform a multifield (MF) classification, which classifies a packet based on multiple fields such as source and destination addresses, source and destination port numbers, type of service (TOS) bits, and protocol ID in the Internet Protocol (IP) header. Then the classified packet will be put into a corresponding queue according to the classification result. Finally, the packet scheduler reorders the output queue to meet different QoS requirements.

The IntServ–RSVP model is not suitable for MANETs due to the resource limitation in MANETs:

1. The amount of state information increases proportionally with the number of flows (the scalability problem, which is also a problem for the current Internet). Keeping flow state information will cost a huge storage and processing overhead for the mobile host, whose storage and computing resources are scarce. Although the scalability problem may not be likely to happen in current MANETs due to their limited bandwidth and relatively small number of flows compared with wired networks, the authors argue that it will occur with the development of fast radio technology and the potential large number of users in the near future.
2. The RSVP signaling packets will contend for bandwidth with the data packets and consume a substantial percentage of bandwidth in MANETs.
3. Every mobile host must perform the processing of admission control, classification, and scheduling. This is a heavy burden for the resource-limited mobile hosts.

6.3.3.2 Differentiated Service (DiffServ)

Differentiated Service (DiffServ) is designed to overcome the difficulty of implementing and deploying IntServ and RSVP in the Internet backbone. DiffServ provides a limited number of aggregated classes to avoid the scalability problem of IntServ. DiffServ defines the layout of the TOS bits in the IP header, called the

"DS field," and a base set of packet-forwarding rules, called "Per-Hop Behavior" (PHB). At the boundary of a network, the boundary routers control the traffic entering the network with classification, marking, policing, and shaping mechanisms. When a data packet enters a DiffServ-enabled domain, a boundary router marks the packet's DS field, and the interior routes along the forwarding path forward the packet based on its DS field. Because the DS field only codes very limited service classes, the processing of the interior routers is very simple and fast. Unlike in IntServ, interior routers in DiffServ do not need to keep per-flow state information. Many services, such as Premium Service, Assured Service, and Olympic Service, can be supported in the DiffServ model. Premium Service is supposed to provide low loss, low delay, low jitter, and end-to-end assured bandwidth service. Assured Service is for applications requiring better reliability than Best Effort Service. Its purpose is to provide guaranteed or at least expected throughput for applications. Furthermore, it is more qualitative oriented than quantitative oriented, and thus is easy to implement. Olympic Service provides three tiers of service—Gold, Silver, and Bronze—with decreasing quality. Supporting Premium Service is almost impossible in the current dynamic MANET environment.

DiffServ may be a possible solution to the MANET QoS model because it is lightweight in interior routers. In addition, it provides Assured Service, which is a feasible service context in a MANET. However, because DiffServ is designed for fixed wire networks, one still faces some challenges to implement DiffServ in MANETs. First, it is ambiguous as to what the boundary routers in MANETs are. Intuitively, the source nodes play the role of boundary routers. Other nodes along the forwarding paths from sources to destinations are interior nodes. But every node should have functionality as both boundary router and interior router because the source nodes cannot be predefined. This arouses a heavy storage cost in every host. Second, the concept of a Service Level Agreement (SLA) in the Internet does not exist in MANETs. The SLA is a kind of contract between a customer and its Internet Service Provider (ISP) that specifies the forwarding services the customer should receive. In the Internet, a customer must have an SLA with its ISP to receive Differentiated Services. The SLA is indispensable because it includes the whole or partial traffic-conditioning rules. Traffic conditioners are placed at the ingress nodes where the traffic originates; they are responsible for re-marking the traffic streams, discarding or shaping packets according to the traffic profile, which describes the temporal properties of a traffic stream such as rate and burst size. How to make an SLA in MANETs is difficult because there is no obvious scheme for the mobile nodes to negotiate the traffic rules.

6.3.3.3 Flexible QoS Model for Mobile Ad Hoc Network (MANET) (FQMM)

A FQMM is proposed. It considers the characteristics of MANETs and tries to take advantage of both the per-flow service granularity in IntServ and the service differentiation in DiffServ. As in DiffServ, three kinds of nodes (ingress, interior, and egress nodes) are defined in FQMM. An ingress node is a mobile node that sends data. Interior nodes are the nodes that forward data for other nodes. An egress node is a destination node. Note that the role of a mobile node is adaptively changing based on its position and the network traffic. The provisioning in FQMM, which is used to determine and allocate the resources at various mobile nodes, is a hybrid scheme of per-flow provisioning as in IntServ and per-class provisioning as in DiffServ. FQMM tries to preserve the per-flow granularity for a small portion of traffic in MANET, given that a large amount of the traffic belongs to per aggregate of flows, that is, per-class granularity.

FQMM is the first attempt at proposing a QoS model for MANETs. However, some problems still need to be solved. First, how many sessions could be served by per-flow granularity? Without an explicit control on the number of services with per-flow granularity, the scalability problem still exists. Second, just as in DiffServ, the interior nodes forward packets according to a certain PHB that is labeled in the DS field.

6.3.4 QoS Frameworks

Framework is a complete system to provide required QoS.

6.3.4.1 INSIGNIA Framework

The INSIGNIA QoS framework allows packet audio, video, and real-time data applications to specify their maximum and minimum bandwidth needs, and plays a central role in resource allocation, restoration control, and session adaptation between communicating mobile hosts. Based on availability of end-to-end bandwidth, QoS mechanisms attempt to provide assurances in support of adaptive services. To support adaptive service, the INSIGNIA QoS framework establishes and maintains reservations for continuous media flows and microflows. To support these communication services, the INSIGNIA QoS framework comprises the following architectural components, as illustrated in Figure 6.4:

- In-band signaling, which establishes, restores, adapts, and tears down adaptive services between source-destination pairs. Flow restoration algorithms respond to dynamic route changes, and adaptation algorithms respond to changes in available bandwidth. Based on an in-band signaling approach that explicitly carries control information in the IP packet header, flows and sessions can be rapidly established, restored, adapted, and released in response to wireless impairments and topology changes.
- Admission control, which is responsible for allocating bandwidth to flows based on the maximum and minimum bandwidth (i.e., base and enhanced QoS)

requested. Once resources have been allocated, they are periodically refreshed by a soft-state mechanism through the reception of data packets. Admission control testing is based on the measured channel capacity and utilization and the requested bandwidth. To keep the signaling protocol simple and light-weight, new reservation requests do not impact existing reservations.

■ Packet forwarding, which classifies incoming packets and forwards them to the appropriate module (viz., routing, signaling, local applications, and packet-scheduling modules). Signaling messages are processed by the in-band signaling, and data packets are delivered locally or forwarded to the packet-scheduling module for transmission to the next hop.

■ Routing protocol, which dynamically tracks changes in ad hoc network topol-ogy making the routing table visible to the nodes' packet-forwarding engine. The QoS framework assumes the availability of a generic set of MANET routing protocols that can be plugged into the architecture. The QoS frame-work assumes that the routing protocol provides new routes, either proactively or on demand, in the case of topology changes.

■ Packet scheduling, which responds to location-dependent channel conditions when scheduling packets in wireless networks. A wide variety of schedul-ing disciplines can be used to realize the packet-scheduling module and the service model. Currently, we have implemented a weighted round-robin service discipline based on an implementation of a deficit round-robin that has been extended to provide compensation in the case of location-dependent channel conditions between mobile nodes.

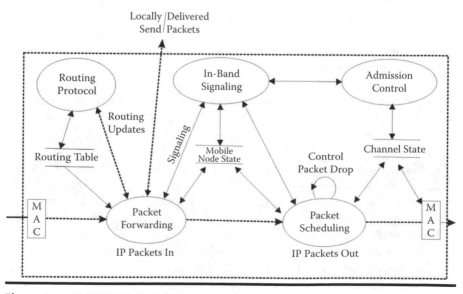

Figure 6.4   INSIGNIA QoS framework.

■ MAC, which provides quality of service–driven access to the shared wireless media for adaptive wireless and best-effort services. The INSIGNIA QoS framework is designed to be transparent to any underlying media access control protocols and is positioned to operate over multiple-link layer technologies at the IP layer. However, the performance of the framework is strongly coupled to the provisioning of QoS support provided by specific medium access controllers.

6.3.4.2 INSIGNIA Signaling System

The INSIGNIA signaling system plays an important role in establishing, adapting, restoring, and terminating end-to-end reservations. In what follows, we describe the INSIGNIA in-band signaling approach. The signaling system is designed to be lightweight in terms of the amount of bandwidth consumed for network control and to be capable of reacting to fast network dynamics such as rapid host mobility, wireless link degradation, and intermittent session connectivity. We discuss first the protocol command and then the protocol mechanisms that implement them.

6.3.4.3 INSIGNIA Protocol Commands

Protocol commands are encoded using the IP option field and include service indicator, payload type, bandwidth indicator, and bandwidth request field, as illustrated in Figure 6.5. By adopting an INSIGNIA IP option in each IP packet header, the complexity of supporting packet encapsulation inside the network is avoided. These protocol commands support the signaling algorithms including flow reservation, restoration, and adaptation mechanisms. The protocol commands drive the state operations of the protocol. Figure 6.4 presents a simplified view of the finite state machines for a source host, intermediate router, and destination hosts. These three state machines capture the major event and actions and the resulting state transitions. We use these state machines to illustrate the dynamics of the INSIGNIA signaling system.

SERVICE TYPE	PAYLOAD INDICATOR	BANDWIDTH INDICATOR	BANDWIDTH REQUEST	
RES/BE	BQ/EQ	BW_IND	MAX	MIN
1 bit	1 bit	8 bits	16 bits	

Figure 6.5 INSIGNIA IP option.

6.3.4.3.1 Service Mode

When a source node wants to establish a fast reservation to a destination node, it sets the reservation mode to RES in the INSIGNIA IP option and sends the packets toward the destination host. On reception of a RES packet, intermediate nodes (functioning as routers) execute admission control to accept or deny the request. When a node accepts a request, resources are committed and subsequent packets are scheduled accordingly. In contrast, if the reservation is denied, packets are treated as Best-Effort (BE) mode packets.

In the case where a RES packet is received and no resources have been allocated, the admission controller attempts to make a new reservation. This condition commonly occurs when flows are rerouted during the lifetime of an ongoing session due to host mobility. When the destination receives an RES packet, it sends a QoS report to the source node indicating that an end-to-end reservation has been established, and transitions its internal state from best-effort to reservation state.

The service mode indicates the level of service assurances requested in support of the adaptive service. The interpretation of the service mode, which indicates an RES or BE packet, is dependent on the payload type and bandwidth indicator. A packet with the service mode set to RES and the bandwidth indictor set to MAX or MIN is attempting to set up max-reserved or min-reserved services, respectively. The bandwidth requirements of the flow are carried in the bandwidth request field, as illustrated in Figure 6.5. An RES packet may be degraded to BE service in the case of rerouting or insufficient resource availability along the new or existing route. Note that a BE packet requires no resource reservation to be made.

The IP option also carries an indication of the payload type, which identifies whether the packet is a base QoS (BQ) or enhanced QoS (EQ) packet. Using the packet state (service mode–payload type–bandwidth indicator), one can determine what component of the flow is degraded. Reception of a BE/EQ/MIN or RES/BQ/MIN packet indicates that the enhanced QoS packets have been degraded to best-effort service. By monitoring the packet state, the destination node can issue scaling and drop commands to the source based on the destination state machine.

As shown in Figure 6.4, the source, intermediate, and destination state machines support two reservation substates:

- Max-reserved mode, which provides reservation for a flow's base QoS and enhanced QoS packets. This type of service requires successful end-to-end reservation to meet a flow's maximum bandwidth needs (e.g., RES/EQ/MAX).
- Min-reserved mode, which provides reservation for the base QoS and best-effort delivery for the enhanced QoS components (if these exist). This service mode typically occurs when max-reserved flows experience degradation in the network. For example, max-reserved flows may encounter mobile hosts and nodes that lack resources to support both the base QoS and the enhanced QoS components, resulting in the degradation of enhanced QoS packets to best-effort delivery (e.g., BE/EQ/MIN).

6.3.4.3.2 Bandwidth Request

The bandwidth request allows a source to specify its Maximum (MAX) and Minimum (MIN) bandwidth requirements for adaptive services. This assumes that the source has selected the RES service mode. A source may also simply specify a minimum or a maximum bandwidth requirement. For adaptive services, the base QoS (min-reserved service) is supported by the minimum bandwidth, whereas the maximum bandwidth supports the delivery of the base and enhanced QoS (max-reserved service) between source–destination pairs. Flows are represented as having minimum and maximum bandwidth requirements. This characterization is commonly used for multiresolution traffic (e.g., MPEG audio and video), adaptive real-time data that has discrete max-min requirements, and differential services that support prioritization of aggregated data in the Internet.

6.3.4.3.3 Payload Type

The payload field indicates the type of packet being transported. INSIGNIA supports two types of payload, BQ and EQ, which are reserved via distributed end-to-end admission control and resource reservation. The semantics of the adaptive services are related to the payload type and available resources (e.g., enhanced QoS requires that maximum bandwidth requirements can be met along the path between a source–destination pair). The semantics of the base and enhanced QoS are application specific. They can represent a simple prioritization scheme between packets, differential services, or self-contained packet streams associated with multiresolution flows. The adaptation process may force adaptive flows to degrade when insufficient resources are available to support the maximum bandwidth along the existing path or during restoration when the new path has insufficient resources. For example, if there is only sufficient bandwidth to meet the minimum bandwidth requirement needs of the base QoS, enhanced QoS packets are degraded to best-effort packets at bottleneck nodes by simply flipping the service mode of EQ packets from RES to BE. When a downstream node detects degraded packets, they release any resources that may have previously been allocated to support the transport of enhanced QoS packets. The adaptation process is also capable of scaling up flows by taking advantage of any additional bandwidth availability that may be encountered along a new or existing path. In this case, a flow could be "scaled up" from min-reserved to max-reserved mode delivery.

6.3.4.3.4 Bandwidth Indicator

A bandwidth indicator plays an important role during reservation setup and adaptation. During reservation establishment, the bandwidth indicator indicates the resource availability at intermediate nodes along the path between a source and destination nodes. Reception of a setup request packet with the bandwidth indicator bit set to MAX indicates that all nodes en route have sufficient resources to support the maximum bandwidth requested (i.e., max-reserved mode). In contrast, a bandwidth indicator set to MIN implies that at least one of the intermediate nodes between the source and destination is a bottleneck node and that insufficient bandwidth is available to meet the maximum bandwidth requirement; that is, only min-reserved mode delivery can be supported. In this case, adaptation algorithms at the destination can trigger the signaling protocol to release any overallocated resources between the source and bottleneck node by issuing a "drop" command to the source node. A bandwidth indicator set to MIN does, however, indicate that the mobile ad hoc network can support the minimum requested bandwidth (i.e., min-reserved mode). The bandwidth indicator is also utilized during the adaptation of ongoing sessions in this manner. The adaptation mechanism resident at the destination host continuously monitors the bandwidth indicator to determine if the additional bandwidth is available to support the better service quality.

6.3.5 INSIGNIA Protocol Operations

In what follows, we provide an overview of the main protocol mechanisms and state machines for the source, intermediate router, and destination nodes as illustrated in Figure 6.4. The key signaling components include reservation establishment, QoS reporting, soft-state management, flow restoration, and flow adaptation.

6.3.5.1 Reservation Establishment

To establish adaptive flows, source nodes initiate reservations by setting the appropriate field in the IP option in data messages before forwarding reservation request packets on toward destination nodes. A reservation request packet is characterized as having the service mode set to RES, payload set to BQ/EQ, and bandwidth indicator set to MAX/MIN and valid bandwidth requirements. Reservation packets traverse intermediate nodes executing admission control modules, allocating resources, and establishing flow state at all intermediate nodes between source–destination pairs, as illustrated in Figure 6.6. A source node continues to send reservation packets until the destination node completes the reservation setup phase by informing the source node of the status of the flow establishment phase using QoS reporting as shown in Figure 6.6.

The establishment of an adaptive flow is illustrated in Figure 6.6. A source node (M_s) requests maximum resource allocation, and node M_1 performs admission control upon reception of the reservation packet. Resources are allocated if available, and the reservation packet is forwarded to the next node M_2. This process is repeated on a hop-by-hop basis until the reservation packet reaches the destination mobile

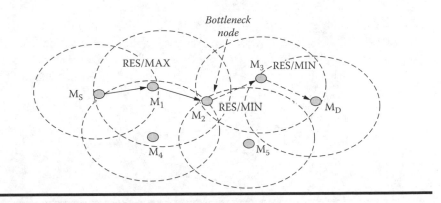

Figure 6.6 Adaptive service flow reservation.

M_D. The destination node determines the resource allocation status by checking the packet state (i.e., service mode, payload type, and bandwidth indicator). The QoS-reporting mechanism is used to inform the source node of the reservation status en route. As far as the destination node is concerned, the reservation phase is complete on reception of the first RES packet. From the example shown in Figure 6.6, we see that only the minimum bandwidth is supported between M_2 and M_3 and that subsequent nodes receiving the request packet avoid allocating resources for the maximum.

When a reservation is received at the destination node, the signaling module checks the flow establishment status. The status of the reservation phase is determined by inspecting the IP option field service mode, which should be set to RES. If the bandwidth indication is set to MAX, this implies that all nodes between a source–destination pair have successfully allocated resources to meet the base and enhanced bandwidth needs in support of the max-reserved mode. On the other hand, if the bandwidth indication is set to MIN, this indicates that only the base QoS can be currently supported (i.e., min-reserved mode). In this case, all reservation packets with a payload of EQ received at a destination will have their service level flipped from RES to BE by the bottleneck node. As a result, "partial reservations" will exist between the source and bottleneck node.

In the case of partial reservations, resources remain reserved between the source and the bottleneck node until explicitly released. Release of partial reserved resources can be initiated by the source based on feedback during the reservation phase or as part of the adaptation process where the destination can issue a scale-down or drop command to a source node. This will have the effect of clearing any partial reservation. An application may choose not to deallocate partial reservation, hedging that bandwidth will become available at the bottleneck node allowing for a full end-to-end reservation to be made in due course.

Note that if a reservation has been established for the maximum reserved state and a RES/BQ/MIN packet is persistently received in this substate, then the state machine determines that the enhanced QoS packets have been degraded and transitions to

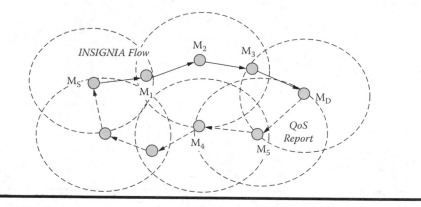

Figure 6.7　QoS reporting.

the minimum reserved state in anticipation of scaling back up. Degradation of this sort can occur at the intermediate node due to insufficient resources to support a new reservation, or an ongoing flow is degraded due to rerouting or insufficient resource availability on the new or existing path. The state information maintained at the destination can decode which of these conditions occurred.

6.3.5.2　QoS Reporting

QoS reporting is used to inform source nodes of the ongoing status of flows. Destination nodes actively monitor ongoing flows inspecting status information (e.g., bandwidth indication) and measuring the delivered QoS (viz., packet loss, delay, and throughput). QoS reports are also sent to source nodes for completing the reservation phase and, on a periodic basis, for managing end-to-end adaptations. QoS reports do not have to travel on the reverse path toward the source. Typically, they will take an alternate route through the ad hoc network, as illustrated in Figure 6.7. Although the QoS reports are basically generated periodically according to the applications' sensitivity to the service quality, QoS reports are sent immediately when required (i.e., typically actions related to adaptation).

In the case where only the BQ packets can be supported, as is the case with the min-reserved mode, the signaling system at the source "flips" the service mode of the BQ packets from RES to BE with all "degraded" packets sent as a best effort. Any partial reservations that may exist between a source and destination nodes automatically time out after "flipping" the state variable in the EQ packets. Because there is a lack of EQ packets with the RES bit set at intermediate routers, any associated resources are released (e.g., between M_S and M_2 in Figure 6.6), allowing other competing flows to contend for these resources. In a similar fashion, QoS reports are also used as part of the ongoing adaptation process that responds to mobility and resources change in the mobile ad hoc network.

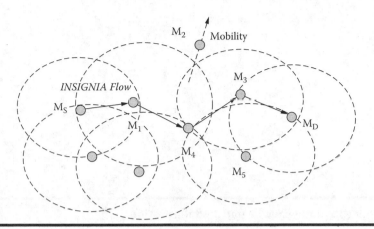

Figure 6.8 Rerouting and restoration.

6.3.5.3 Flow Restoration

Flows are often rerouted within the lifetime of ongoing sessions due to host mobility. The goal of flow restoration is to reestablish reservation as quickly and efficiently as possible. Rerouting active flows involves the routing protocol (to determine a new route), admission control, and resources reservation for nodes that belong to a new path. Restoration procedures also call for the removal of the old flow state at nodes along the old paths. In an ideal case, the restoration of flows can be accomplished within the duration of a few consecutive packets given that an alternative route is cached. We call this type of restoration "immediate restoration." If no alternative route is cached, the performance of the restoration algorithm is coupled to the speed at which the routing protocols can discover a new path.

As illustrated in Figure 6.8, network dynamics trigger rerouting and service degradation. In this example, mobile host M_2 moves out of radio contact and connectivity is lost in Figure 6.8. The forwarding router node, M_1, interacts with the routing protocol and forwards packets along a new route. The signaling system at intermediate router M_4 receives packets and inspects its flow soft-state table. If a reservation does not exist for newly arriving packets, then the signaling module invokes admission control and attempts to allocate resources for the flow. Note that when a rerouted packet arrives at node M_3, the forwarding engine detects that a reservation exists and treats the packet as any other packet with a reservation. In other words, the packets are routed back onto the existing path where a reservation is still present. Soft-state timers ensure that the flow state is still intact at M_3 and that states along the old path (i.e., the mobile host M_2) are removed in an efficient manner.

When an adaptive flow is rerouted to a node where resources are unavailable, the flow is degraded to best-effort service. Subsequently, downstream nodes receiving these degraded packets do not attempt to allocate resources or refresh the

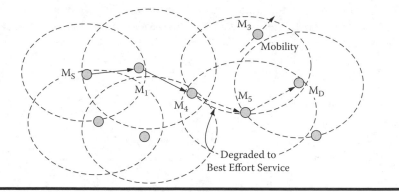

Figure 6.9   Rerouting and degradation illustration.

reservation state associated with the flow. In this instance, the state associated with a flow is timed out and resources are deallocated. A reservation may be restored if the resources free up at a "bottleneck" node (e.g., mobile node M_4 in Figure 6.9) or further rerouting may allow the restoration to complete. We call this type of restoration "degraded restoration." A flow may remain degraded for the duration of the session, never restoring, which is described as "permanent degradation." The enhanced QoS component of an adaptive flow may be degraded to best-effort service (i.e., min-reserved mode) during the flow restoration process if the nodes along the new path can only support the minimum bandwidth requirement. If the degradation of enhanced QoS packets persists, it may cause service disruption and trigger the destination mobile node to invoke its adaptation procedure to "scale down" or "drop" packets rather than live with degraded quality. Adaptation mechanisms located at destination nodes are capable of responding to changes in network resource availability through "scale-down," "scale-up," and "drop" actions in response to network conditions.

During the restoration process, the INSIGNIA framework does not favor rerouted flows over existing flows in favor of some fairness criteria (e.g., by forcing existing flows to scale down to their minimum requirements to allow rerouted or new flows to be admitted). In this sense, INSIGNIA avoids the introduction of additional service fluctuations of existing flows in support of the restoration of rerouted flows. As a result of this policy, admission control simply rejects or scales down any rerouted flows when insufficient resources are available along a new path.

Three types of restoration are supported by the INSIGNIA QoS framework:

- ◼ Immediate restoration, which occurs when a rerouted flow immediately recovers its original reservation; that is, a max-reserved mode flow is immediately restored as a max-reserved mode flow, and a min-reserved mode flow as a min-reserved mode flow.

■ Degraded restoration, which occurs when a rerouted flow is degraded for a period (T) before it recovers to its original reservation. Two forms of degraded restoration can occur:

1. A max-reserved mode flow operates at min-reserved mode or best-effort mode, and eventually recovers its original max-reserved mode service after some interval.

2. A min-reserved mode flow operates at best-effort mode and eventually recovers its original min-reserved mode service after some interval.

■ Permanent degradation, which occurs when the rerouted flow never recovers its original reservation.

Figure 6.9 illustrates the topology changes that occur after rerouting based on the initial topology shown in Figure 6.8. After rerouting link M_4, M_5 can only support best-effort services. This type of restoration represents either a degraded restoration or a permanent degradation. In this scenario, the destination node clears the partial reservation between mobile nodes M_5 and M_4 by issuing a drop adaptation command to the source. The process of restoration can be immediate or delayed. Adaptation is application specific, where the application can choose to respond to the network conditions and the delivered QoS.

6.3.5.4 Flow Adaptation

The INSIGNIA QoS framework actively monitors network dynamics and adapts flows in response to observed changes based on user-supplied adaptation policy. Flow reception quality is monitored at the destination node and based on application-specific adaptation policy; actions are taken to adapt flows under certain observed conditions. Action taken is conditional on what is programmed into the adaptation policy by the user. For example, an adaptation policy could be to maintain the service level under degraded conditions or scale down adaptive flows to their base QoS in response to degraded conditions; other policy aspects could be to always scale up adaptive flows whenever resources are available. The application is free to program its own adaptation policy, which is executed by INSIGNIA through interaction of the destination and source nodes.

INSIGNIA provides a set of adaptation levels that can be selected. Typically, an adaptive flow operates with both its base and enhanced components being transported with resource reservation. Scaling flows down depending on the adaptation policy selected. The flow can be scaled down to its base QoS delivering enhanced QoS packets in a best-effort mode, hence releasing any partial reservation that may exist. On the other hand, the destination can issue a drop command to the source to drop enhanced QoS packets (i.e., the source stops transmitting enhanced QoS packets). Further levels of scaling can force the base and enhanced QoS packets to be fully transported in best-effort mode. In both cases, the time scale over which

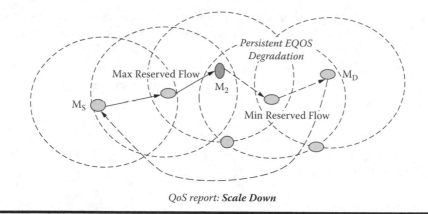

*QoS report: **Scale Down***

Figure 6.10 Scaling-down adaptation.

the adaptation actions occur is dependent on the application itself. These scaling actions could be instantaneous or based on a low-pass filter operation.

During restoration of the flow state, admission control and resource reservation are invoked. This can lead to changes in a flow's observed quality at the destination mode in terms of having to both scale down flows in response to observed resource bottlenecks along the new path and scale up flows when additional resources are made available along the new path.

The INSIGNIA signaling system supports three adaptation commands that are sent from the destination host to the source using QoS reports:

■ Scale-down command, which requests a source node to send its enhanced QoS packets as its best effort or its enhanced QoS and base QoS as its best effort
■ Drop command, which requests a source node to drop its enhanced QoS packets or enhanced and base QoS packets (where the term "drop" means the source node stops transmitting these packets)
■ Scale-up command, which requests a source node to initiate a reservation for its base or enhanced service quality

The scale-down, drop, and scale-up actions are driven by adaptation policy implemented at the destination, as illustrated in Figure 6.10 and Figure 6.11. Note that preference is given to base over enhanced QoS components in the event that reserved packets have to be degraded to the best-effort mode at bottleneck nodes, as illustrated in the figure. The scale-down command is issued when the degradation of enhanced QoS packets persists. This action forces source nodes to send the enhanced QoS packets as best-effort packets, thereby effectively removing any partial reservations that may exist, as illustrated in the figure. A drop command is issued only when a destination node determines that degraded packets render insufficient quality. It is up to the applications to decide whether the reception of

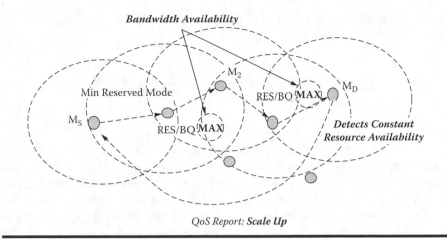

*QoS Report: **Scale Up***

Figure 6.11 Scaling-up adaptation.

degraded packets is acceptable or not and take the appropriate action. An adaptation policy handler at the destination is free to issue scale-down commands or, in the case of persistent degradation (possibly including best-effort delivery of both the base and enhanced QoS components), to terminate the session.

Mobility results in the release of resources along old paths, and session dynamics result in additional resources becoming available along existing paths when sessions terminate. These released resources help other source–destination pairs support higher levels of quality for their sessions, assuming they share a common path with that of the released resources. In such a case, the signaling system sets the bandwidth indication in the packet's INSIGNIA IP option field to indicate to adaptation handlers (located at the receiver) that sufficient resources may be available to support the delivery of base and enhanced QoS. The signaling system uses the bandwidth indication field to inform the destination host of the availability of new network resources should they become available along an existing path. Bottleneck nodes set the bandwidth indicator to MIN when enhanced QoS packets are scaled back in response to degraded conditions. Because each packet carries the max-min bandwidth requirements of each flow, bottleneck nodes can update a packet's bandwidth indicator in the event that resources become available to meet enhanced QoS needs. If all nodes along a path have resources to support enhanced QoS, then the bandwidth indicator received at the destination will indicate MAX in the bandwidth indicator field. This does not imply that a reservation has been made or that a reservation could be made with a 100 percent assurance. Rather, it indicates to the source node that a reservation may be possible and that at the time the bandwidth indicator bit was set, resources were available. To initiate the reservation for the enhanced QoS adaptation, handlers send "scale-up" commands to their respective source nodes. In this sense the bandwidth indicator represents a good resource hint that additional service quality is possible. All messaging between source–destination pairs in support of scaling or dropping flow components is achieved using QoS reports.

6.3.6 Intelligent Optimization Self-Regulated Adjustment (INORA)

INORA is a QoS support mechanism that makes use of the INSIGNIA in-band signaling and TORA routing protocol for MANETs. INORA represents a QoS-signaling approach in a loosely coupled kind of manner. The idea is based upon the property of TORA to provide multiple routes between a given source and destination. Although INSIGNIA does not take any help from the network with regard to redirecting the flow along routes that are able to provide the required QoS guarantees, INORA gives feedback to the routing protocol on a per-hop basis to direct the flow along the route that is able to satisfy the QoS requirements of the flow. Beyond a doubt, the concept of "loosely coupling" QoS signaling and routing is a very promising approach, and the shortcomings of INORA mostly are the shortcomings of INSIGNIA. However, the interface for signaling to access routing should be as generic as possible to guarantee portability.

TORA operates by creating a Directed Acyclic Graph (DAG) rooted at the destination. The DAG is extremely useful in this scheme because it provides multiple routes from the source to the destination. It used this routing structure to direct the flow through routes that are able to provide the resources for the flow according to the QoS requirements of the flow. INORA can be classified into two schemes: the (1) coarse-feedback scheme and (2) fine-feedback scheme.

6.3.6.1 Coarse-Feedback Scheme

The operations of the coarse-feedback scheme of INORA are illustrated through the following example. Consider a QoS flow being initiated with node 1 as the source and node 5 as the destination.

1. Let the DAG created by TORA be as illustrated in Figure 6.12. Let $1 \rightarrow 2 \rightarrow 3 \rightarrow 4 \rightarrow 5$ be the path chosen by the TORA routing protocol.
2. INSIGNIA tries to establish soft-state reservations for the QoS flow along the path. Node 4 is the first node at which admission control for the flow fails. Node 4 sends an out-of-band Admission Control Failure (ACF) message to its previous hop (node 3; see Figure 6.13).
3. Node 3 realizes that the next hop, node 4, is not good for the current flow and reroutes the flow through another downstream neighbor (node 6) provided by TORA (Figure 6.14).
4. If node 6 is able to admit the flow, the flow gets the required reservations all along the path. The new path would be $1 \rightarrow 2 \rightarrow 3 \rightarrow 6 \rightarrow 5$ (Figure 6.14).

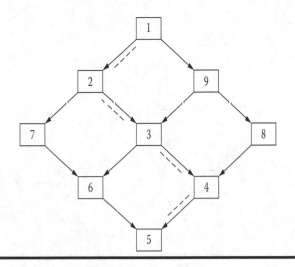

Figure 6.12   INORA coarse feedback.

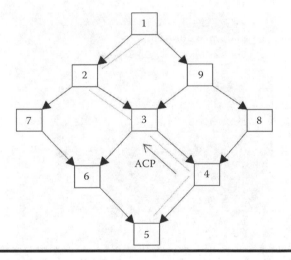

Figure 6.13   INORA coarse feedback: node 4 sends out-of-band ACF to previous hop (node 3).

5. If node 6 is unable to admit the flow, it sends an ACF message to node 3 (its previous hop; see Figure 6.15).

6. Node 3 realizes that it has exhausted all the downstream neighbors that it was provided by TORA. So, it sends an ACF message to its previous hop (node 2), indicating that none of its downstream neighbors can accommodate the flow (Figure 6.16).

7. Node 2 now tries with its other downstream neighbors for the possibility of a path that can give the required reservations to the flow (Figure 6.17).

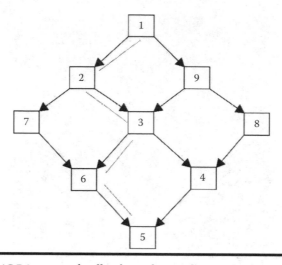

Figure 6.14 INORA coarse feedback: node 3 redirects the flow to node 6.

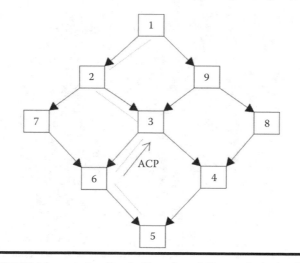

Figure 6.15 INORA coarse feedback: if node 6 fails to admit the flow, it sends an ACF message to node 3.

Note that as a result of this scheme, it is possible that different flows between the same source and destination pair can take different routes, so that to go from node 1 to node 5, flow 1 takes the path $1 \rightarrow 2 \rightarrow 3 \rightarrow 4 \rightarrow 5$ and flow 2 takes the path $3 \rightarrow 6 \rightarrow 5$.

Although INORA is trying to find a good route for the flow following the ACF at an intermediate node, the packets are transmitted as BE packets from the source to the destination. It should also be noted that there is no interruption in the transmission of a flow that has not been able to find a route in which resources have been reserved all the way from the source to the destination.

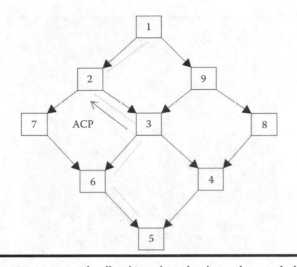

Figure 6.16   INORA coarse feedback: node 3, having exhausted all its next hops, sends an ACF message to its previous hop, node 2.

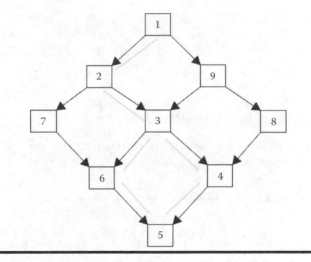

Figure 6.17   INORA coarse feedback: different flows between the same source–destination pair can take different routes.

Because of the nature of the DAG, INORA tries to get a route which satisfies QoS requirements locally. When this fails, the search for a route which satisfies the QoS requirement becomes more global. In the worst case, we would have searched the entire DAG for a QoS route. The state introduced in the nodes due to this search is soft. So, there is no overhead in maintaining the state. This search, which goes on in the background, does not affect the delivery of the packets of the flow. Also, the scope of search for the routes is the DAG. INORA only chooses an appropriate route from the set of routes given by TORA. It does not trigger any route-querying

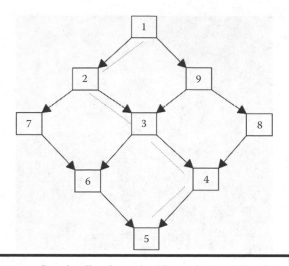

Figure 6.18 INORA fine feedback: node 3 has admitted the flow with class l but is not able to give the bandwidth class that node 2 (previous hop) is able to give, say, m(m > l).

mechanism to find routes which will satisfy the QoS requirements. The philosophy of INORA is that it tries to find better routes for a flow while the transmission of the flow continues without interruption.

6.3.7 Class-Based Fine Feedback Scheme

In this scheme, we divide the (BWmin; BWmax) interval into *N* classes, where BWmin is the minimum bandwidth required by a QoS flow and BWmax is the maximum bandwidth required by the QoS flow. The IP options field in the IP header, which carries the INSIGNIA information, now carries an additional class field. This field signifies the amount of bandwidth that has been allocated for the flow along the path. The operation of the protocol is illustrated by the following example.

Consider a QoS flow being initiated with node 1 as the source and node 5 as the destination, with minimum bandwidth requirement BWmin and maximum bandwidth requirement BWmax. Let the flow be admitted with class (m < N) m at node 1.

1. Let the DAG created by TORA be as shown in Figure 6.18. Let $1 \rightarrow 2 \rightarrow 3 \rightarrow 4 \rightarrow 5$ be the path chosen by the routing protocol.
2. INSIGNIA tries to establish soft-state reservations for the QoS flow along the path.
3. Node 2 is able to admit the flow with class m as was requested by its previous upstream hop, node 1.

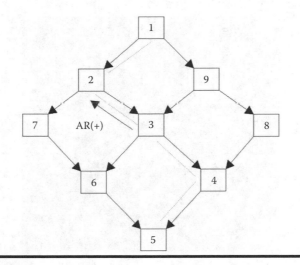

Figure 6.19 INORA fine feedback: node 3 sends Admission Report (AR) l to node 2.

4. Suppose that node 3 has admitted the flow with class l, but has not been able to allocate the bandwidth of class m(l < m), as requested by its previous hop, node 2.

5. Node 3 now sends an Admission Report (AR) message AR (l) to the upstream previous hop (node 2), indicating its ability to give class l bandwidth to the flow (Figure 6.19).

6. Node 2 splits the flow in the ratio of l to (m − l) and forwards the flow to node 3 and node 7, respectively, in that ratio. This means that the flow of class m has been split into two flows of class l and (m − l) and is forwarded to nodes 3 and 7, respectively (Figure 6.20).

7. Suppose that node 7 is unable to give class (m − l) as requested by the upstream previous hop, node 2, but is only able to give class n (< m − l). Node 7 sends an AR message (n) to the upstream previous hop, node 2 (Figure 6.21).

8. Node 2, realizing that its downstream neighbors have been unable to give the class m, which it had requested, informs of its ability to give a class (l + n < m) l + n by sending an AR to its previous hop, node 1 (Figure 6.22).

9. Node 1 now tries to find another downstream neighbor, which might be able to accommodate the flow with class m − (l + n) (Figure 6.23).

Note that when a node is unable to admit a flow, either because it cannot allocate the minimum bandwidth BWmin required by the flow or due to congestion at a node, it sends ACF messages as in the coarse-feedback scheme. So, the fine-feedback scheme includes the features of the coarse-feedback scheme. The fine-feedback scheme, like the coarse-feedback scheme, first tries to search for a QoS route, which can give the requested bandwidth class locally. The search becomes more global if it is not able to find locally the QoS route that gives the required

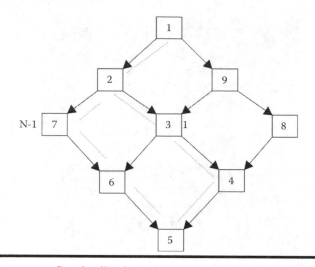

Figure 6.20 INORA fine feedback: node 2 splits the flow between the next hops, 7 and 3, in the ratio (m – l) to l, respectively.

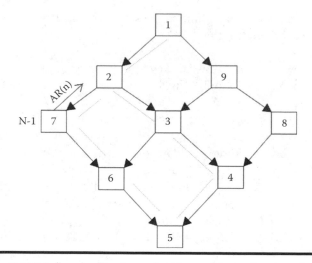

Figure 6.21 INORA fine feedback: node 7 is unable to give (m – l), but only n < (m – l). It sends AR (n) upstream.

cumulative class. Also, it is worth noting that a single flow can get split, and the packets can take different routes from the source to the destination (Figure 6.14). This can result in packets being received out of order at the destination. The real-time applications with QoS requirements typically use Real-Time Transport Protocol (RTP) as the transport protocol. RTP does reordering of the packets. If TCP is used as the transport protocol, packets arriving out of sequence can trigger TCP's congestion avoidance mechanisms. The effect of out-of-order delivery on TCP has to be further investigated.

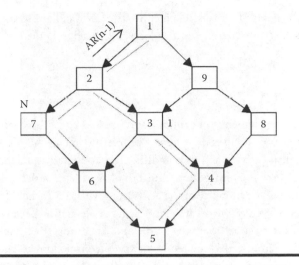

Figure 6.22   INORA fine feedback: node 2 sends AR (n + l), indicating the bandwidth that it can support.

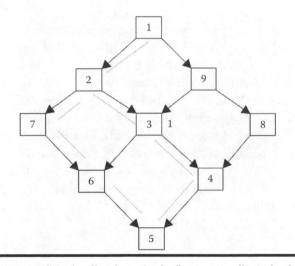

Figure 6.23   INORA fine feedback: a single flow gets split and takes different paths to the destination.

We evaluate the performance of the INORA schemes by observing the end-to-end delay of the packets and the control message overhead. We see that the INORA coarse-feedback scheme has lesser average delay than INSIGNIA and TORA operating without feedback. The INORA fine-feedback scheme performs better than the INORA coarse-feedback scheme. This is because the INORA feedback schemes try to find paths which can allocate the requested bandwidth reservations to the QoS flows. The fine-feedback scheme does this in a much finer-grained manner when compared to the coarse-feedback scheme. So, the authors have found the fine-feedback scheme to perform better than the coarse-feedback scheme.

6.4 Summary

In this chapter, the challenge of bringing QoS support to mobile ad hoc networks has been discussed. It focuses on different problems that are highly related to each other and that deal with some common difficulties, which include mobility, limited bandwidth and power consumption, and broadcast characteristics of radio transmission. Because of the mobility of the hosts in MANETs, an established path may break and the reserved resources may not be available again. Because mobile hosts could move in an arbitrary way, the traditional meaning that some performance metrics must be guaranteed once a request is accepted is no longer true in MANETs. In addition, the topology changes could also affect the available bandwidth. How to minimize the influence of the hosts' movement should be the first consideration for QoS support in MANETs. Because of the bandwidth and power limitations, the cost on providing QoS should be controlled in a reasonable range.

The objective here is not to design strong but complex QoS mechanisms. Instead, finding a way to support a sort of QoS service with little burden on the mobile hosts will be a main principle in current MANETs. Because of the broadcast characteristics of radio transmission, abundant hidden or exposed terminals may exist, which make reliable transmission and end-to-end delay control very difficult. Because the transmission between neighboring nodes could interfere with each other, the available bandwidth in a mobile host changes with the surrounding environment, such as how many neighboring hosts are contending for the transmission channel, how many neighboring hosts are moving away, or how many nonneighboring nodes are moving near. All these factors will increase the complexity of QoS support in the MANET environment. As a whole, the above difficulties constitute the main challenges for QoS support in MANETs.

6.5 Problems

6.1 Quality of Service is needed at all layers. Justify.
6.2 Describe the challenges and issues involved in providing QoS.
6.3 Give the classifications of QoS solutions.
6.4 Explain different types of QoS models with suitable illustrations.
6.5 Discuss the INSIGNIA framework QoS model.
6.6 Write a note on QoS reporting.
6.7 Describe the INORA framework with a suitable example.

Bibliography

G. Ahn, A. Campbell, A. Veres, and L. Sun, SWAN: Service differentiation in stateless wireless ad hoc networks, in Proceedings of IEEE INFOCOM, 2002.

D. Black, Differentiated services and tunnels, RFC2983, 2000.

R. Braden, Resource reservation protocol: Version 1 message processing rules, RFC2209, 1997.

R. Braden, D. Clark, and S. Shenker, Integrated services in the Internet architecture: An overview, technical report 1633, 1994.

R. Braden, L. Zhang, S. Berson, S. Herzog, and S. Jamin, Resource reservation protocol (RSVP): Version 1 functional specification, RFC2205.

R. Braden, D. Estrin, S. Berson, S. Herzog, and D. Zappala, The design of the RSVP protocol.

K. Chaing, C. H. Xiao, and W. K. G Seah, A quality of service model for ad hoc wireless networks.

K. Chandran, S. Raghunathan, S. Venkatesan, and R. Prakash, A feedback-based scheme for improving TCP performance in ad hoc wireless networks, in IEEE *Personal Communications*, Feb. 2001.

L. Chen and W. Heinzelman, End-to-end congestion control for best-effort transmission, in Proceedings of Wireless Networking Symposium, October 2003.

L. Chen and W. Heinzelman, QoS-aware routing based on bandwidth estimation in mobile ad hoc networks, in Technical Report of University of Rochester, NY, 2003.

B-G. Chun and M. Baker, Evaluation of packet scheduling algorithm in mobile ad hoc networks, *Mobile Computing and Communications Review*, 2002, http://www.cs.berkeley.edu/~bgchun/manetsched-mc2r02.pdf.

A. Conta, A proposal for the IPV6 flow label specification, 2001, http://www.ietf.org/proceedings/02mar/I-D/draft-ietf-ipv6-flow-label-00.txt.

Z-Y. Demetrios, A glance at quality of services in mobile ad hoc networks, in Seminar in Mobile Ad Hoc Networks, 2001.

D. Dharmaraju, A. Roy-Chowdhury, P. Hovareshti, and J. S. Baras, INORA: A unified signaling and routing mechanism for QoS support in mobile ad hoc networks, in Proceedings of the International Conference on Parallel Processing Workshops, 2002.

Y. Ge, T. Kunz, and L. Lamont, Quality of service routing in ad-hoc networks using OLSR, in Proceedings of the 36th Hawaii International Conference on System Science, 2003.

K. Geihs, Analysis of adaptation strategies for mobile QoS-aware applications, in Workshop on Modeling Analysis and Simulation of Wireless and Mobile Systems, 2002.

M. Greis, RSVP/NS: An implementation of RSVP for the network simulator ns-2.

R. Hinden and S. Deering, Internet protocol, version 6 specification, RFC1883, 1995.

Imaq: An integrated mobile ad hoc QoS framework, http://cairo.cs.uiuc.edu/adhoc/.

IP QoS support in the Internet backbone, technical report, Alcatel.

A. Kassler, T. Guenkova-Luy, D. Mandator, and P. Schoo, Enabling mobile heterogeneous networking environments with end-to-end user perceived QoS, technical report, IST project IST-2000-28584 MIND.

C.R. Lin and M. Gerla, MACA/PR: An asynchronous multimedia multihop wireless network, in Proceedings of IEEE INFOCOM '97.

A. López, J. Manner, A. Mihailovic, H. Velayos, E. Hepworth, and Y. Khouaja, A study on QoS provision for IP-based radio access networks, IST project IST-1999-10050 BRAIN, 1999.

S. Mangold, S. Choi, P. May, O. Klein, G. Hiertz, and L. Stibor, IEEE 802.11e wireless LAN for quality of service, European Wireless, 2002, http://citeseer.ist.psu.edu/537394.html.

J. Manner, A. López, A. Mihailovic, H. Velayos, E. Hepworth, and Y. Khouaja, Evaluation of mobility and QoS interaction, in *Computer Network*, Elsevier Science, Vol. 38, Feb. 2002, pp. 137–163.

M. Mirhakkak, N. Schult, and D. Thomson, A new approach for providing quality of service in a dynamic network environment, MILICOM 2000, Vol. 2, pp. 1020–1025.

B. Moon and H. Aghvami, RSVP extensions for real-time services in wireless mobile networks, *IEEE Communication*, Vol. 39, Dec. 2001, pp. 52–59.

The network simulator ns-2, http://www.isi.edu/nsnam/ns/.

B. Shrader, A proposed definition of "ad hoc network," 2002.

P. Sinha, R. Sivakumar, and V. Bhargavan, CEDAR: A core-extraction distributed ad hoc routing algorithm, in IEEE INFOCOM'99, New York.

A. K. Talukdar, B. R. Badrinath, and A. Acharya, MRSVP: A resource reservation protocol for an integrated services network with mobile hosts, *Wireless Networks*, 7(1):5–19, 2001.

P. P. White, RSVP and integrated services: A tutorial, 1997.

K. Wu and J. Harms, QoS support in mobile ad hoc networks, Crossing Boundaries, *GSA Journal of University of Alberta*, 1(2):92–106, Nov. 2001.

H. Xiao, K. C. Chua, and W. K. G. Seah, A quality of service model for ad hoc wireless networks, in *Handbook of ad hoc wireless networks*, CRC Press, Boca Raton, FL, 2002.

H. Xiao, K. C. Chua, and W. S. A. Lo, On service prioritization in mobile ad-hoc networks, in Proceedings of the ICC, 2001.

J. Xue, Adaptive QoS-supporting architecture for real-time application in wireless IP networks, master's thesis, 2002, http://www.inf.ethz.ch/personal/jxue/.

Q. Xue and A. Ganz, Ad hoc QoS on-demand routing (AQOR) in mobile ad hoc networks, *Journal of Parallel and Distributed Computing*, Feb. 2003.

X. Zhang, S-B. Lee, G-S. Ahn, and A. T. Campbell, Evaluation of the insignia signaling system, Proceedings of the IFIP-TCG/European Commission International Conference on High Performance Networking, May 2000, pp. 311–324.

X. Zhang, S-B. Lee, G-S. Ahn, and A. T. Campbell, INSIGNIA: An IP-based quality of service framework for mobile ad hoc networks, 1999.

L. Zhang, S. Deering, D. Estrin, S. Shenker, and D. Zappala, RSVP: A new resource reservation protocol, *IEEE Network*, Sept. 1993.

Chapter 7

Energy Management Systems in Ad Hoc Wireless Networks

7.1 Introduction

In the last few years, there has been an explosive growth of interest in mobile computing, as well as in delivering World Wide Web content and streaming traffic to radio devices. There is a huge potential market for providing palmtops, laptops, and personal communication systems with access to airline schedules, weather forecasts, or location-dependent information, to name just a few. Wireless ad hoc networks can be used to provide radio devices with such services, anywhere and anytime. Ad hoc networks enable users to spontaneously form a dynamic communication system. They allow users to access the services offered by the fixed network through multihop communications, without requiring infrastructure in the user proximity. However, to offer high-quality and low-cost services to ad hoc network nodes, several technical challenges still need to be addressed. First, wireless networks are plagued by scarcity of communication bandwidth; therefore, a key issue is to satisfy user requests with minimum service delay. Second, because network nodes have limited energy resources, the energy expended for transferring information across the network has to be minimized. Ad hoc wireless networks are constrained by limited battery power, which makes energy management an important issue.

Energy management deals with

■ Efficient battery management
■ Transmission power management

7.1.1 Why Energy Management Is Needed in Ad Hoc Networks

In ad hoc wireless networks, mobile computation devices are usually battery powered. A limited energy budget constrains the computation and communication capacity of each device. Energy resources and computation workloads have different distributions within the network. Some mobile devices have spare energy. Devices that expend all their energy can only be recharged when they leave the network. Therefore, it is beneficial to redistribute spare energy resources to satisfy unevenly distributed workloads in the network.

In wireless networks, the ratio of computation energy consumption to communication energy consumption varies in a wide range, depending on application type. In some applications, for example microsensor networks, communication dominates energy consumption. In other application domains and applications, for example simulation, classification, artificial intelligence, target detection, handwriting recognition, and voice recognition, computation energy consumption generally dominates communication energy consumption. If devices with excess computation-intensive tasks can, for a fee, transfer these tasks to devices with spare energy and time, both buyer and seller devices benefit; sellers may use their earnings to buy energy in the future.

The main reasons for energy management in ad hoc networks are as follows:

Limited energy reserve: The ad hoc networks have limited energy reserve. The improvement in battery technologies is very slow as compared to the advances in the field of mobile computing and communication.

Difficulties in replacing the batteries: In situations like battlefields, natural disasters such as earthquakes, and so forth, it is very difficult to replace and recharge the batteries. Thus, in such situations, the conservation of energy is very important.

Lack of central coordination: Because an ad hoc network is a distributed network and there is no central coordinator, some of the nodes in the multihop routing should act as a relay node. If there is heavy relay traffic, this leads to more power consumption at the respective relay node.

Constraints on the battery source: The weight of the nodes may increase with the weight of the battery at that node. If the weight of the battery is decreased, that in turn will lead to less power of the battery and thus decrease the life span of the battery. Thus, energy management techniques must deal with this issue; in addition to reducing the size of the battery, they must utilize the energy resources in the best possible way.

Selection of optimal transmission power: The increase in the transmission power increases the consumption of the battery charge. Because the transmission power decides the reachability of the nodes, an optimal transmission power decreases the interference between nodes, and that in turn increases the number of simultaneous transmissions.

7.1.2 Classification of Energy Management Schemes

To increase the life of an ad hoc wireless network, we should have a better understanding of the capabilities and limitations of the energy resources of the nodes. A longer lifetime of the node can be achieved by increasing the battery capacity. Increasing the capacity of the battery at the nodes can be achieved by either battery management, which concerns the internal characteristics of the battery, or power management, which deals with utilizing the battery capacity to the maximum possible extent.

Figure 7.1 provides an overview of the classification of the battery management system. The battery management system can be divided into three categories:

■ Battery management system
■ Transmission power management
■ System power management

The system power management can be further subdivided into the following categories:

■ Device management schemes
■ Processor power management schemes

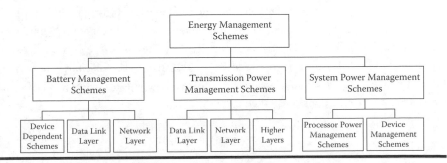

Figure 7.1   Classification of energy management schemes.

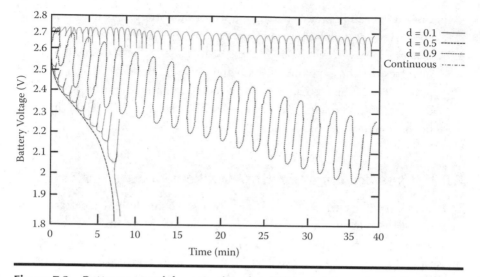

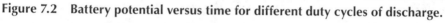

Figure 7.2   Battery potential versus time for different duty cycles of discharge.

7.1.2.1   Battery Management Schemes

Portable user terminals for mobile communications must rely on limited battery power for their operation. It has been shown that energy savings in portable devices can be sought at different layers of the wireless protocol stack not necessarily at the device or circuit level alone. Also, several studies characterizing the battery discharge behavior have shown that "pulsed discharge" performs better than continuous discharge. Particularly, the battery can recharge itself (i.e., recover the potential) if left idle after discharge. This "recharge phenomenon" is illustrated in Figure 7.2. The figure shows the battery potential (in volts) as a function of time with continuous discharge and pulsed discharge. The parameter represents the duty cycle of the pulsed discharge pattern (i.e., d = 0.1 means 10 percent discharge time followed by 90 percent relaxation time, and so on). A 2.76-volt (V) Lithium ion battery with a cutoff voltage of 1.9 V is considered. It is assumed that the battery ceases to deliver power once the voltage drops below the cutoff voltage. In other words, the time taken for the battery to fall below the cutoff voltage is the battery life. The slope of battery discharge is determined by the discharge current density. The larger the value of this current density is, the steeper the discharge slope will be, and hence the time taken to reach the cutoff voltage will be lesser. In Figure 7.2 the discharge current density is taken to be 5 A/m².

$$\text{Current density} = 5 \text{ A/m}^2$$

$$\text{Cutoff voltage} = 1.9 \text{ V}$$

From Figure 7.2 it can be seen that if the battery is discharged continuously, it takes about 7.5 minutes to reach the cutoff voltage. On the other hand, if the battery is discharged in pulsed mode, the battery recovers the voltage during the relaxation periods and it takes more time to reach the cutoff voltage. For example, if the duty cycle of the pulsed discharge is 50 percent (i.e., d = 0.5), it takes more than 40 minutes to reach the cutoff voltage. Also, because the duty cycle is 0.5, the total "on time" of the battery is more than 20 minutes, which is about three times the on time in continuous discharge mode. As the discharge current density is taken to be the same in both continuous and pulsed discharge modes, this essentially means that the battery can deliver energy for a longer duration. This recharge effect advantage in pulsed mode can be exploited for improved energy efficiency in packet communications in wireless mobile devices. Through simulations, it can be shown that the recharge phenomenon can be exploited for battery life gain through suitable traffic-shaping algorithms and battery-level sensed routing strategies. The battery can be assumed as a server with finite capacity, and the arriving packets at the mobile terminal as the customers to be served. Each transmitted packet consumes energy proportional to the packet size, transmission bit rate, wireless link design, and so forth. The server (battery) should be allowed to go on vacation for a calibrated amount of time, essentially allowing idle times for the battery to recharge itself. By doing so, the number of customers served can be increased (in other words, battery life can be increased). We derive expressions for the number of customers served and the average delay for an M/G1/1 queuing system without and with server vacations. It can be shown that allowing forced vacations during busy periods helps to increase battery life. However, because the customers (packets) have to wait in the queue when the server goes on vacation, the battery life gain will come at the expense of increased delay performance of the packets. It would be of interest to guarantee specified packet delay performances in practical systems. Hence, to achieve bounded delay performance, we further propose an algorithm that will exploit recharge phenomenon *when packet delay constraints* are imposed.

Battery-powered electronic systems, and the integrated circuits within them, account for a large and rapidly growing revenue segment for the computer, electronics, and semiconductor industries. For instance, the revenue from wireless voice or data handsets is expected to exceed that from personal computers (PCs) in the near future, and the use of wireless Internet access is expected to overtake fixed Internet access in the next few years. For battery-powered systems, the battery life directly impacts the system's utility and the duration and extent of its mobility. The battery life of a system is determined by the capacity of the energy source (i.e., battery), and the energy drawn by the rest of the system. Improvements in semiconductor fabrication and wireless communication technologies promise to enable advances in ubiquitous information access and manipulation (anytime, anywhere computing and communications). Unfortunately, projected improvements in the capacity of batteries (5 to 10 percent compound annual growth rate) are much slower than what is needed to support the increasing complexity, functionality, and performance

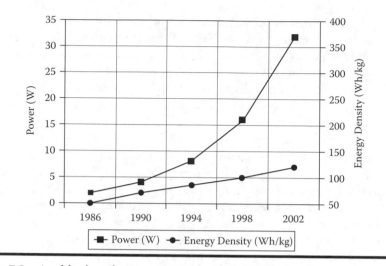

Figure 7.3 A widening "battery gap," due to rapidly increasing power requirements and slowly improving battery technology.

of the systems they power. Figure 7.3 illustrates a widening "battery gap" between
trends in processor power consumption and improvements in battery capacity.

Bridging this gap is a challenge that system designers must face for the foresee-
able future. The need to improve battery life has, in large part, driven the research
and development of low-power design techniques for electronic circuits and systems.
Low-power design techniques are successful in reducing the energy that is drawn
from the battery, and hence improve battery life to some extent. However, truly
maximizing battery life requires an understanding of both the source of energy and
the system that consumes it. It has been shown that the amount of energy that can
be supplied by a given battery varies significantly, depending on how the energy is
drawn. "Battery-driven system design," which refers to the process of designing a
system with careful consideration of the battery and its characteristics, promises
to provide further improvements in battery life beyond what can be achieved by
conventional low-power design technologies.

7.1.3 Overview of Battery Technologies

In this subsection, we describe battery technologies that have been developed over
the last two decades to meet the increasing demand for smaller, lighter, higher
capacity rechargeable batteries for portable appliances. When comparing different battery technologies, several considerations arise. These include energy density
(charge stored per unit weight of the battery), cycle life (the number of discharge and
charge cycles prior to battery disposal), environmental impact, safety, cost, available supply voltage, and charge and discharge characteristics. Figure 7.4 illustrates

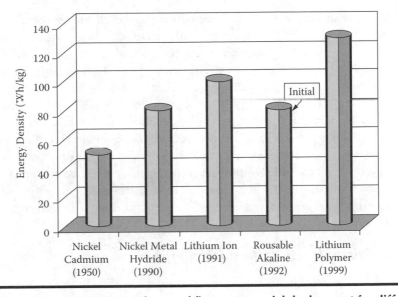

Figure 7.4   Energy density and year of first commercial deployment for different battery technologies.

the development of rechargeable battery technologies, and compares them in terms of typical energy density with the most popular rechargeable battery technologies.

The most popular rechargeable battery technologies for portable electronic appliances include the following:

Nickel cadmium (Ni-Cd): This is a mature technology, and has been successfully used for several decades to develop rechargeable batteries for portable electronic devices. Its advantages include low cost and high discharge rates. Although Ni-Cd technology has been losing ground in recent years owing to its low energy density and toxicity, it is still used in low-cost applications like portable radios and CD/tape players.

Nickel metal-hydride (Ni-MH): These batteries have been in widespread use in recent years for powering laptop computers. They have roughly twice the energy density of Ni-Cd batteries. However, they have shorter cycle life, are more expensive, and are inefficient at high rates of discharge.

Lithium ion: This is the fastest growing battery technology today, with significantly higher energy densities and a cycle life about twice that of Ni-MH batteries. Lithium ion batteries are more sensitive to characteristics of the discharge current, are more expensive than Ni-MH batteries, and can be unsafe when improperly used. However, longer lifetimes have made them the most popular battery choice for notebooks, PDAs, and cellular phones.

Reusable alkaline: Although disposable alkaline batteries have been used for many years, reusable alkaline manganese technology has developed as a low-cost alternative in which energy density and cycle life are compromised. Although the initial energy density of reusable alkaline batteries is higher than that of Ni-Cd batteries, it has been found to decrease rapidly with cycle life. For instance, after ten cycles, a 50 percent reduction is commonly observed, and after 50 cycles, a 75 percent reduction in energy density is commonly observed.

Lithium polymer: This emerging technology enables ultrathin batteries (less than one millimeter thickness), and is expected to suit the needs of lightweight, next-generation portable computing and communication devices. Additionally, they are expected to improve over current lithium ion technology in terms of energy density and safety. However, these batteries are currently expensive to manufacture and face challenges in internal thermal management.

7.1.4 Principles of Battery Discharge

The basic components of a battery are shown in Figure 7.5, through the example of a lithium–thionyl chloride cell. The battery consists of an anode (lithium), a cathode (carbon), and an electrolyte. The electrolyte separates the two electrodes and provides a mechanism for the transfer of charge between them. During battery discharge, oxidation at the anode (Li) results in the generation of (1) electrons, which flow through the external circuit; and (2) positively charged ions (Li+), which, by diffusion, move through the electrolyte ($SOCl_2$) toward the cathode. Reduction reactions occur at available reaction sites in the cathode, generating negatively charged ions (Cl–), which combine with the positive ions (Li+) to generate an insoluble compound (LiCl) that gets deposited on the cathode. Sites where the compound is deposited become inactive, making them unavailable for further use. As discharge proceeds, more and more reaction sites are made unavailable, eventually leading to a state of complete discharge.

A battery is characterized by the open-circuit voltage (VOC), that is, the initial potential of a fully charged battery under no-load conditions, and the cut-off voltage (Vcut) at which the battery is considered discharged. Battery capacity can

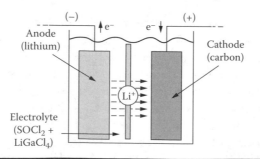

Figure 7.5   Basic structure of a lithium–thionyl chloride battery.

be expressed in three ways. The *theoretical capacity* of a battery is based on the amount of energy stored in the battery, and is an upper bound of the total energy that can be extracted in practice. The *standard capacity* of a battery is the energy that can be extracted when it is discharged under standard load conditions, which are specified by the manufacturer. For example, a typical lithium ion battery may have a standard capacity of 500 milliamp hours when discharged at a constant current of 125 milliamperes, at 25°C. The *actual capacity* of a battery is the amount of energy that the battery delivers under a given load, and is usually used (along with battery life) as a metric to judge the battery efficiency of the load system. A battery-efficient system is one where the discharge profile characteristics result in improved actual capacity. Although the actual capacity may exceed the standard capacity of the battery, it cannot exceed its theoretical capacity.

7.1.5   Impact of Discharge Characteristics on Battery Capacity

Two important effects that make battery performance sensitive to the profile of the discharge current are (1) *rate capacity effects*, which are due to a dependency between the actual capacity of a battery and the magnitude of the discharge current; and (2) *recovery effects*, which are due to the recovery of charge during idle periods (when no charge is drawn). Both these phenomena can significantly affect the capacity and the lifetime of a battery.

7.1.5.1   Rate Capacity Effects

The lifetime of a battery depends on the availability and reachability of active reaction sites in the cathode. During periods of discharge where the rate of discharge is low (the current drawn is small), the distribution of inactive reaction sites throughout the material of the cathode is more or less uniform. However, if the rate of discharge is high (a large current is drawn), reductions occur only at the outer surface of the cathode. This results in the surface of the cathode being coated with an insoluble compound, preventing access to many active internal reaction sites. Consequently, the battery is declared discharged even though many active cathode sites remain unutilized, effectively decreasing the total capacity of the battery. The effect of this phenomenon is a dependency between battery capacity and the rate at which it is discharged.

7.1.5.2   Recovery Effects

Besides the availability of active reaction sites in the cathode, the availability of charged ions (Li+) is also a factor that determines the amount of energy that can be delivered by a battery. When no current is drawn, the concentration of positively charged ions (Li+) is uniform at the electrode–electrolyte interface. When current is drawn from the battery, positively charged ions are consumed at the cathode–electrolyte interface, and replaced by new ions that diffuse from the anode through the electrolyte. When the current drawn is sufficiently large, the rate of diffusion fails to keep up with the rate at which ions are consumed at the cathode. As a result, the concentration of positively charged ions decreases near the cathode and increases near the anode, degrading the battery's output voltage. However, if the battery is allowed to idle for a period of time, the concentration gradient decreases (due to diffusion), and charge recovery takes place at the cathode. As a result, the capacity and lifetime of the battery increase. To summarize, the amount of energy that can be extracted from a battery, and consequently its lifetime, are sensitive to (1) the magnitude of the discharge current, and (2) the presence of idle times in the discharge profile. We next turn our attention to battery models that have been developed to capture such battery characteristics.

7.1.6 Battery Modeling

Battery models capture the characteristics of real-life batteries, and can be used to predict their behavior under various conditions of charge and discharge. Battery models are useful tools for a battery-driven system design approach, because they enable analysis of the discharge behavior of the battery under different design choices (e.g., system architectures and power management policies), without resorting to time-consuming (and expensive) prototyping and measurement for each alternative. In the following discussion, we consider several different battery models. We classify these approaches into (1) analytical models, (2) electrical circuit models, (3) stochastic models, and (4) electrochemical models. Additionally, there are several other important factors that characterize a battery model, each of which may also serve as a basis for classification. For example, battery models could be classified on the basis of (1) the types of load current supported (e.g., constant load versus variable load), (2) supported battery technologies (Lithium ion, Ni-Cd, etc.), (3) the set of battery effects captured by the model (e.g., rate capacity effects, recovery effects, and thermal effects), (4) computational efficiency, and (5) accuracy in predicting real-life battery behavior.

7.1.6.1 Analytical Models

Several battery models have been developed where analytical expressions are formulated to calculate actual battery capacity and lifetime using discharge current values, operating environment characteristics, and physical properties of the battery as parameters. One of the simplest and earliest analytical models is Peukert's formula, which expresses the nonlinear relationship between the battery capacity and the rate of discharge. Peukert's formula states that the actual capacity Q of a battery is given by

$$Q = \frac{K}{I^\alpha}$$

where I is the load current (assumed to be constant); k is a constant capturing electrochemical properties, physical construction, and operating environment of the battery; and K is a constant that captures the rate capacity effect. Recently, sophisticated analytical models have been proposed, where the capacity is expressed in terms of more complex functions of discharge current, temperature, and several constants. Various techniques are used to determine the exact form of the function. For example, regression analysis is used to estimate coefficients in an expression based on a Weibull failure model. This is a constant load model that captures rate capacity and thermal effects. The model described is a variable load model that uses empirical equations to capture the rate capacity effect. It can be used to predict variation in capacity for current discharge profiles that follow different probability density functions. A variable load model uses laws of chemical kinetics to derive the mathematical expression, and uses statistical techniques to estimate the parameters. In summary, analytical battery models can include both constant-load and variable-load models. All of these models capture rate capacity effects, and some capture thermal effects, though none address recovery due to relaxation during idle periods. These models are flexible, and can be easily configured for specific batteries. They are computationally efficient, requiring evaluation of simple analytical expressions.

7.1.6.2   Electrical Circuit Models

We next consider a class of techniques that model battery discharge using an equivalent electrical circuit. Most of these approaches are based on constructing a Small Profile Intelligent Coverage Element (SPICE) model of a coupled network to represent the battery. The charge stored in the battery is modeled using a capacitor, and voltage across the capacitor is used to represent the output voltage. The discharge process is modeled by continuously applying corrections to both the charge stored in the capacitor and the output voltage. For example, to model the rate capacity effect, a voltage source is used to subtract from the voltage across the capacitor. The discharge current indexes into a lookup table to control this voltage source. Further corrections account for (1) the cycle life of the battery (the number of times the battery has been charged and discharged), (2) internal resistance (which may vary with time), and (3) thermal characteristics. Electrical circuit models are variable-load models that are capable of modeling rate capacity and thermal effects. However, none of the known circuit models account for recovery effects. This class of battery models lends itself to easy simulation using existing circuit simulators, and has been used in practice to analyze many common battery technologies.

7.1.6.3 Stochastic Models

A stochastic model of a battery is represented by a finite number of charge units, and the discharge behavior of the battery is modeled using a discrete-time transient stochastic process. As the stochastic process evolves over time (which is divided into a sequence of equally sized slots), the state of the battery is tracked by the number of remaining units of charge. In each time slot, the average discharge current is measured and used to determine the number of charge units consumed. If this average is nonzero, the number of charged units drained is obtained from a look-up table (or a plot) that contains rate capacity data. However, if the slot is an idle slot (i.e., no current is drawn), then the battery recovers a certain number of charge units. The exact number of charge units recovered is modeled using a decreasing exponential probability density function, which is based on (1) the state of charge of the battery, and (2) coefficients, which depend on the specific battery as well as the discharge characteristics. Over time, the battery steps through a sequence of states, from a state of full charge to either (1) a state where the cutoff voltage is reached, or (2) a state where the theoretical capacity is exhausted. The stochastic model described above can account for variable loads, and currently has been used to model lithium ion batteries. It accounts for both rate capacity and recovery effects. However, it does not take into account thermal effects. Its accuracy stems from the fact that accurate battery specifications are used to construct rate capacity data, and from the detailed analysis performed for the exact formulation of the probability density functions. Moreover, its computation requirements are modest, enabling it to be employed in system-level simulations.

7.1.6.4 Electrochemical Models

This includes the class of battery models that directly consider the electrochemical processes, thermodynamic processes, and physical construction when modeling discharge of the battery. These models are significantly more detailed (and hence more accurate) than any of the previously described approaches. Some models describe concentrated solution theory, and porous electrode theory is used to construct a set of differential equations for a specific battery. The model has a large number of parameters, including electrode geometries, concentration of the electrolyte, diffusion coefficients, transfer coefficients, and reaction rate constants.

Numerical techniques are used to solve the equations to predict battery capacity under different load conditions. Battery models that belong to this class are very closely tied to specific batteries.

7.1.7 Battery-Driven System Design

In this section, we describe various approaches to the design of battery-efficient systems. The first class of techniques that we consider is based on optimizing the system architecture itself. These include battery-driven policies for frequency scaling, task scheduling, voltage scaling, and power management. The second class of techniques we consider addresses multibattery systems through battery-scheduling techniques aimed at optimizing the discharge of constituent batteries. Finally, we describe techniques that enable the battery-efficient operation of portable appliances in a wireless network by making use of new network protocols and traffic-shaping techniques.

7.1.7.1 Battery-Efficient System Architectures

In this subsection, we present approaches to the battery-efficient design of hardware–software system architectures.

7.1.7.1.1 Frequency Scaling

Frequency scaling is a common approach to reducing the average power consumption, and Central Processing Unit (CPU) frequency scaling for battery-powered computers is examined in terms of its impact on battery life, system performance, and power consumption. A commonly used history-based policy (not optimized for batteries) for CPU frequency scaling dynamically calculates the CPU frequency for the next time interval based on (1) run- and idle-time percentages of the previous time interval, and (2) work left over, in case the CPU frequency was too slow in the previous time interval. A larger percentage of idle time in the previous time interval results in decreasing the CPU clock frequency by a small constant for the next time interval while not going below a lower bound. A larger percentage of run-time results in increasing CPU frequency by a small constant while not going beyond the CPU's maximum clock frequency. It is shown that by using the above history-based frequency-scaling policy, important factors like nonideal battery behavior are neglected. Hence, the lower bound on CPU clock frequency is redefined to maximize a metric combining battery capacity, performance, and power. In summary, frequency-scaling approaches use information from a battery model to vary the clock frequency of system components dynamically at run time. Because they also use workload characteristics (run- and idle-time percentages) and models of system power and performance, these approaches can be used to ensure efficient use of the battery without significantly compromising system performance.

7.1.7.1.2 Battery-Aware Task Scheduling

Battery-aware static scheduling algorithms have been developed for real-time systems that aim at improving the system's current discharge profile. In this approach, a static schedule is generated from a given task graph using standard scheduling algorithms. The schedule is then subjected to a series of postprocessing stages, wherein transformations are applied to optimize the corresponding discharge current profile for the battery. The transformations are applied to the schedule in a two-step manner. First, the initial schedule is transformed using "global shifting," which reduces current peak consumption and increases flexibility in the schedule. This serves as a high-quality initial solution for the second step, which is a set of local transformations to the schedule. These local transformations attempt to change the position of scheduled events (e.g., interchange, shift forward, and shift backward) to minimize a cost function that captures delay and the actual energy drawn from the battery. This approach explicitly tailors the current discharge profile of the system to meet battery characteristics. However, its applicability is limited to systems that can be statically scheduled.

7.1.7.1.3 Supply Voltage Scaling

We next describe recent approaches that apply battery-driven techniques toward scaling the supply voltage of the system. An analytical technique is presented to optimally select V_{dd} (the supply voltage) to find the best trade-off between battery capacity and performance. The objective is to find a value of V_{dd} that minimizes the "battery discharge-delay" product, which is given by the product of the actual charge drained from the battery and the delay of the circuit for a given task. To achieve this, the battery discharge-delay product is mathematically expressed in terms of V_{dd} and other known (or measurable) parameters. The analysis uses (1) an analytical model of the battery, (2) the distribution of the discharge current profile, and (3) a model for Complementary Metal-Oxide Semiconductor (CMOS) circuit delay in terms of V_{dd} and V_{th}. Note that this approach does not attempt to modify the shape of the current discharge profile, either statically or dynamically. Rather, if the current discharge profile (or distribution) can be statically determined, this technique can be used to select a constant supply voltage to jointly optimize circuit delay and battery life. The approach applies to statically scheduled real-time systems, and is based on reallocating slack times for tasks to enable supply voltage scaling. In this technique, an initial static schedule, which satisfies all performance constraints, is subjected to a set of global and local transformations to (1) shape the current discharge profile with a knowledge of the battery rate capacity characteristics, and (2) facilitate reallocation of the amount of slack assigned to each task. The slack available for various tasks is then exploited by voltage scaling, with the objective of flattening the discharge profile while meeting all real-time constraints.

7.1.7.1.4 Dynamic Power Management

Although system-level power management is in itself a well-researched area, recent research has proposed new power management schemes that specifically target battery efficiency rather than average power. The battery-driven system-level power management technique is based on a policy that controls the operation state of the system according to the state of charge of the battery. As a case study, a battery-powered digital audiorecorder is considered. The aim of the power management policy is to provide a graceful degradation in audio quality as the battery approaches a state of complete discharge, along with an extension in battery life. To achieve this objective, the dynamic power management policy used is as follows. As long as the battery output voltage is above a certain threshold, the recorder plays high-quality music, which results in a high rate of battery discharge (low battery efficiency). When the battery output voltage falls below a threshold voltage, the policy forces the player to degrade the quality of music, causing the system to discharge the battery at a lower rate (higher battery efficiency). By including feedback from the battery, this work extends standard dynamic power management techniques (e.g., timeouts), which depend only on the energy-consuming components of the system and their workload. These power management policies exploit rate capacity effects to improve battery efficiency of the system when the battery nears a state of complete discharge. This approach can be described as being "reactive," because the battery-driven policies come into play only when the state of charge of the battery falls below a certain threshold.

7.1.7.2 Battery Scheduling and Management

With increasing energy requirements in portable appliances, systems with multiple batteries are no longer uncommon. A typical approach taken in the case of these appliances is to discharge each battery sequentially and completely. However, recent work has shown this policy may not be the best for battery life. In this subsection, we discuss approaches toward efficient management of multibattery systems. In particular, we describe contributions made toward battery scheduling: how to distribute the current demand of a system among a set of batteries. We consider three classes of battery-scheduling techniques. The first class includes static-scheduling techniques (which do not use any run-time information), whereas the second and third classes are dynamic-scheduling techniques that use measurements of battery terminal voltage and discharge current, respectively.

7.1.7.2.1 Static Battery Scheduling

Static battery-scheduling approaches do not make use of any run-time information from either (1) the system discharge profile, or (2) the state of charge of the various batteries. These techniques include the following:

Serial scheduling: In this approach, all the batteries are discharged in sequence—an approach currently adopted in most products that are powered by multiple batteries.

Random scheduling: In this approach, at every discharge interval, a battery is chosen at random from the array of batteries to power the system for a fixed discharge interval.

Round-robin scheduling: In this approach, for each discharge interval, the battery to be used is chosen in a round-robin fashion.

Both random and round-robin scheduling result in an improvement in the lifetime of the system as compared to serial scheduling. That is because these approaches provide individual batteries the opportunity to recover lost capacity during idle periods at times when other batteries are being used. Experiments indicate that the round-robin scheme performs better than the random scheme (when the current demand follows a Poisson distribution). Both these approaches require a selector circuit to switch between batteries at some frequency F_{sw}. The value of F_{sw} directly relates to the discharge interval, the duration for which each battery is continuously discharged without idling. When $F_{sw} = 0$, the batteries are discharged in sequence. However, improvements in battery capacity are observed with increasing F_{sw}, till such time at which diminishing returns are obtained, due to (1) large time constants internal to the batteries, and (2) energy consumed by the switching circuit.

7.1.7.2.2 Terminal Voltage-Based Battery Scheduling

Several scheduling techniques have been suggested that make use of information regarding the state of charge of the battery, estimated from the terminal voltage of the battery. Modified round-robin approaches take into account the output voltage. In these approaches, a fixed round-robin scheme is used till the voltage of one or more batteries falls below some threshold voltage V_{th}. Batteries with $V_{out} < V_{th}$ are disconnected from the load (made "inactive"), which gives them a chance to recover some charge. Meanwhile, round-robin scheduling is applied among the remaining "active" batteries in the pack. Inactive batteries may recover enough charge to reenter the set of active batteries. Finally, when no active batteries are left (i.e., with $V_{out} > V_{th}$), the batteries are again discharged in round-robin fashion. The discharge interval is not fixed (as in the previously described approaches) but is adapted to the state of charge. The policy used is as follows: the battery chosen for discharge is used continuously till the output voltage falls below a certain threshold, at which point it is disconnected from the load. The next battery to be used is chosen based on the output voltages and the past idle times of the various candidate batteries.

7.1.7.2.3 Discharge Current-Based Battery Scheduling

Consider a two-battery system with batteries named A and B. At low rates of discharge, battery A has higher capacity per unit weight than battery B. However, at high rates of discharge, battery B has higher capacity per unit weight than A. In other words, battery B lasts longer under high rates of discharge, whereas A lasts longer under low rates of discharge. The following scheduling policy is used to maximize the lifetime of this two-battery supply: if the rate of discharge is less than a threshold current, battery A is connected to the system; otherwise, battery B is connected. Experiments show that using the two-battery supply with discharge current–based scheduling significantly improves the lifetime of the system, compared to the case where a single large battery of type A or B is used.

7.1.7.3 Battery-Efficient Traffic Shaping and Routing

In a network of battery-operated devices, the architecture of each constituent node is not the only factor that determines battery lifetime. In this case, an important role is played by the network protocols and communication traffic patterns in determining battery efficiency and lifetime. In this subsection, we highlight some recent work in networking protocols and traffic-shaping algorithms for wireless networks that aim at improving battery life of the mobile nodes. The idea is to forcibly interrupt the discharge of the battery whenever the battery charge falls below a certain level, and allow it to recover lost capacity. During these interruptions, incoming packets awaiting processing are buffered. Discharge of the battery resumes when (1) the battery recovers sufficient charge to process at least one request, and (2) there are one or more pending requests. A battery-efficient routing protocol for ad hoc wireless networks, the metric for choosing the best route between a source and destination, is based on (1) the state of charge of the battery-powered nodes along the route, (2) the transmission energy of each node on the route, and (3) penalties due to rate capacity considerations for nodes that transmit at high power levels. The protocol selects routes with low impact on the battery capacity of the constituent nodes, and hence allows recovery of charge to occur on inactive nodes.

7.1.8 Smart Battery Systems

In this section, we briefly describe the Smart Battery System (SBS) standard, which is an emerging industry standard relating to smart batteries and the systems that deploy them. Initially developed by Intel and Duracell, the SBS specifications are currently being developed by an industry consortium called the SBS Implementers Forum. The SBS specification is currently supported by over 80 companies that include battery, semiconductor, software, and system manufacturers, and suppliers of battery electronics. The aim of the SBS forum is to create open standards that enable systems to be aware of and better communicate with the batteries that power them, improve battery efficiency, and promote interoperability between products from battery, software, semiconductor, and system vendors.

The SBS specifications include the following components:

System management bus (SMBus), which defines the protocols for the battery to communicate with other system components. Because the SMBus can also be used as a control bus for other low-speed system communications, the SMBus specifications are defined by an independent forum.

Battery data set, which defines the information that is provided by a smart battery to the system host, smart charger, smart battery system manager, and other system components. Accuracy of the battery data set, which determines the accuracy of the battery information presented by the system to the user, is also covered by the specifications.

Smart battery charger, which enables the charging characteristics to be controlled by the batteries themselves, in contrast to conventional chargers that have fixed charging characteristics hardwired for specific battery chemistries and configurations. Smart battery chargers result in improvements in system safety, usable energy, charging time, and cycle life.

Smart battery system selector, which is used in multibattery systems to select the battery that will actually supply power to the load system. It is also responsible for reporting any changes in the selector state to the system's power management software.

Smart battery system manager, which manages the usage of all the smart batteries in a system. The SBS manager is an alternative to the smart battery selector that provides for added functionality, including the possibility for multiple batteries to simultaneously power the system. It also schedules and controls the charging of multiple batteries, and reports the characteristics of batteries powering the system to the management software.

A typical example of a smart battery system is shown in Figure 7.6 that includes two smart batteries, a smart battery charger, and an SBS manager, which communicate among themselves and with the host system via the SMBus. In the figure, the SBS manager has configured smart battery 1 to power the system while smart battery 2 is charging. In addition, the flow of information relating to critical events (e.g., battery low alarm) and battery data and status requests is shown in the figure. Smart battery systems promise several advantages: (1) better and more accurate sensing of battery data by the system will lead to battery-aware power management policies and hence improved battery life; (2) systems will be able to use batteries of any chemistry type, and new technologies can be directly utilized by existing SBS compatible systems; (3) smart charging will improve the safety and cycle life while reducing charging times; and (4) battery data can be used not only by the

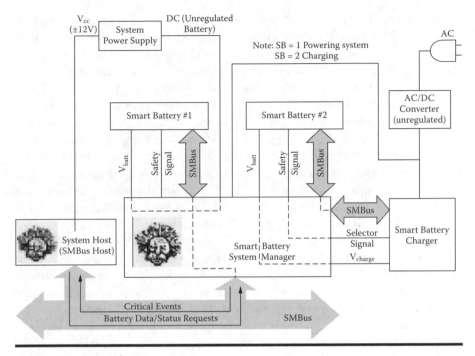

Figure 7.6   Smart battery system.

host system but also by systems across the network, eventually supporting remote diagnosis and repair.

7.2   Energy-Efficient Routing Protocol

Energy efficiency is a major challenge in wireless networks. To facilitate communication, most wireless network devices are portable and battery-powered, and thus operate on an extremely constrained energy budget. However, progress in battery technology shows that only small improvements in battery capacity can be expected in the near future. Furthermore, because recharging or replacing batteries is costly or, under some circumstances, impossible, it is desirable to keep the energy dissipation level of devices low. A mobile ad hoc network is a collection of two or more nodes equipped with wireless communications and networking capabilities without central network control, namely, an infrastructureless mobile network. Energy-efficient design in mobile ad hoc networks (MANETs) is more important and challenging than with other wireless networks. First, due to the absence of an infrastructure, mobile nodes in an ad hoc network must act as routers and join in the process of forwarding packets. Therefore, traffic loads in MANETs are heavier than in other wireless networks with fixed access points or base stations, and thus MANETs have more energy consumption. Second, energy-efficient design needs to consider the trade-offs between different network performance criteria. For example, routing protocols usually try to find the shortest path from sources to destinations. It is possible that some key nodes will overserve the network and have their energy drained quickly, causing the network to be partitioned. Thus simple solutions that only consider power constraints may cause severe performance degradation. Third, no centralized control implies that energy-efficient management in MANETs must be done in a distributed and cooperative manner, which is difficult to achieve.

At the wireless interface, energy consumption in idle mode is only slightly less than in transmit mode and almost equal to that of receive mode. Therefore, it is desirable to build a network protocol that maximizes the time a device is in sleep mode (namely, with the wireless interface off) and also maximizes the number of wireless devices that can be in sleep mode. Many protocols have been proposed to deal with this challenge. In this chapter, a new Energy-Efficient Medium Access Control (EE-MAC) Protocol is proposed. The design is based on the fact that most applications of ad hoc networks are data driven, which means that the sole purpose of forming an ad hoc network is to collect and disperse data. Hence, keeping all network nodes awake is costly and unnecessary when some nodes do not have traffic to carry. The protocol conserves energy by turning on and off the radios of specific nodes in the network. The goal is to reduce energy consumption without significantly reducing network performance. EE-MAC is based on the Institute of Electrical and Electronics Engineers (IEEE) 802.11 and its power-saving mode (PSM), and can provide useful information to the network layer for route discovery.

7.2.1 An Overview of IEEE 802.11 Power-Saving Mode

Power management can achieve great savings in infrastructure networks. All traffic for mobile stations must go through access points, so they are ideal locations to buffer traffic. However, in ad hoc networks, far more of the burden is placed on the sender to ensure that the receiver is active or awake. Receivers must also be more available and cannot sleep for as long as in infrastructure networks. Power management in IEEE 802.11 PSM is based on traffic indication messages. Nodes use Asynchronous Traffic Indication Maps (ATIMs) to notify other nodes to prepare to receive data. All nodes have to wake up periodically to listen for ATIMs and check whether they have packets to receive. The time in PSM can be divided into beacon intervals, and each beacon interval starts with an ATIM window. This window is the period during which nodes must remain active, and no stations are permitted to power down their wireless interface. The ATIM window size is a parameter that can be set. Setting it to 0 means no power management is used. There are four possibilities for a node in terms of ATIMs: the node has transmitted an ATIM, received an ATIM, neither transmitted nor received, or both transmitted and received. Nodes that transmit ATIM frames do not sleep because this indicates an intent to transmit buffered traffic. Nodes to which an ATIM is addressed must also keep awake so

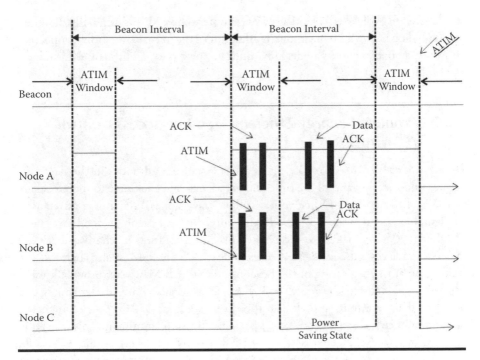

Figure 7.7   IEEE 802.1 1 Power-Saving Mode (PSM) operation.

they can receive data packets from the ATIM sender. A node that both transmits and receives, of course, needs to be active. Thus, only those nodes that neither transmit nor receive an ATIM can go to sleep after the ATIM window. Figure 7.7 illustrates the basic PSM operations.

Nodes A and B have advertised packets in the ATIM window by sending ATIMs and receiving ATIM–acknowledgment (ACK) packets, both of which are subject to the distributed coordination function (DCF) rules described earlier. Therefore, nodes A and B remain awake for the rest of the beacon interval. The transmission of data packets from nodes A and B takes place during the beacon interval. The node that has no packets to transmit can go into the sleep state at the end of the ATIM window if it does not receive an ATIM during the window. In the example in Figure 7.7, node C enters sleep mode after the ATIM window, thus saving energy. All sleeping nodes wake up at the start of the next beacon interval. The beacon and ATIM window sizes can affect the performance of PSM. Because no data packets are transmitted in the ATIM window, overhead in terms of energy consumption and bandwidth is incurred. If we use a small ATIM window to improve energy savings, there may not be enough time to advertise all buffered data packets. Conversely, using a large ATIM window may unnecessarily waste bandwidth and not leave enough time to transmit buffered data. Moreover, PSM also suffers from long packet delivery latency because for each hop that a packet traverses, the packet can be expected to be delayed for at least a beacon period. PSM was originally designed for single-hop networks, which means all nodes in the network are fully connected. However, ad hoc networks are usually multi hopnetworks, and thus PSM is not an ideal solution.

7.2.2 Proposed Energy-Efficient Medium Access Control (EE-MAC) Protocol

The key idea of EE-MAC is to elect master nodes from all nodes in the network. Master nodes stay awake all the time and act as a virtual backbone to route packets in the ad hoc network. Other nodes, called "slave nodes," remain in an energy-efficient mode and wake up periodically only during beacon intervals to check whether they have packets to receive. To be fair, a rotation mechanism between masters and slaves is used. EE-MAC uses some features of PSM, such as periodically waking up at the beginning of the beacon interval. EE-MAC can provide knowledge and guidance to the route lookup process, because only master nodes can be selected along a routing path. On the other hand, EE-MAC requires a mechanism to awaken a sleeping node when packet delivery is imminent. This is usually handled by low-level mechanisms at the MAC or physical layers. In EE-MAC, if a node has been asleep for a while, packets addressed to it are not lost but stored at one of its upstream nodes, usually a master. When the node awakens, the buffered data is sent to it, and this is a PSM feature which is used in our protocol.

7.2.2.1 Design Criteria

We consider the following design criteria.

The protocol must ensure enough master nodes are elected to build the backbone of the network so that every node has at least one master in its vicinity. A collection of masters can be described as a Connected Dominating Set (CDS). All nodes are either a member of the CDS or a direct neighbor of at least one of the members of the CDS. Nodes in the CDS serve as the routing backbone and remain active all the time. All other nodes are slave nodes and can choose to sleep. Because slave nodes do not join in the process of route discovery or packet forwarding, the network's level of connectivity is decreased. To prevent a dramatic decrease in throughput, an acceptable set of masters is required to maintain global connectivity with some redundancy.

The master nodes' election algorithm is based on local information, which is a distributed approach. Each node only consults its local information to determine whether it will become a master. Due to the characteristics of distributed management in ad hoc networks and the two essential requirements, low overhead and fast convergence, the algorithm for finding a CDS should be localized.

The algorithm must have a fair way to rotate masters and slaves to ensure that all nodes share the job of providing global connectivity equally. Overusing some critical nodes will severely decrease the network lifetime of an ad hoc network. Thus, if some alternative nodes appear, masters can step down and give the new nodes a chance to serve as masters to balance node energy consumption.

7.2.2.2 Features of EE-MAC

In EE-MAC, because masters do not operate in power-saving mode and can forward packets all the time, the packet delivery ratio and packet delay can be improved greatly compared to PSM. In this section, we present some features of EE-MAC.

Entering sleep mode earlier: In the original PSM, a node with packets to transmit will send an ATIM frame to the destination, and both source and destination will keep awake in that beacon interval, no matter how many packets need to be transmitted. Although this approach has its advantages, it may result in much higher energy consumption than necessary. For example, if a source only has one packet pending, it must still be on during the entire beacon period to deal with this packet. To avoid this, we add the number of remaining data packets at the sender into every data packet sent to the destination. This information allows the destination to know when it has received all pending packets for it. When the source and destination have sent or received all their packets, they can enter sleep mode until the beginning of the next beacon interval.

Priority processing of packets to slaves: When nodes are trying to send packets, they first deal with those to be sent to slave nodes. After transmitting all packets to slave nodes, packets between masters can be sent. With this approach, slaves can stay as long as possible in sleep mode.

Prolonging the sleep period for slaves: In EE-MAC, most packets are forwarded by masters, and packet routing via slaves is kept to a minimum. To take advantage of this, each slave uses history information to decide on its sleep time. When a node observes two consecutive beacon intervals without any packets addressed to it, it will decide to sleep through the next beacon interval. At the same time, this slave's master stores this information because failure to get an ACK does not guarantee a broken link. If the master does not know a slave's situation, it just buffers the packets to that slave. Only when the master does not hear from a neighboring slave for two consecutive beacon intervals does it discard these packets.

Additional MAC layer control: Nodes in an ad hoc network may move randomly. Thus, to quickly adapt to network topology changes, a node informs its neighbors of its status, master or slave, by using the power management bit in the MAC header. Because the MAC header can be heard anywhere in the network, including request-to-send and clear-to-send packets, this information will help neighbors to know each other's situation.

7.2.2.3 Performance

We use the following metrics to evaluate the network performance. Note that these metrics differ from those used by others. Because our main concern is energy efficiency, energy level is given a higher weight than connectivity.

Data packet delivery ratio: The data packet delivery ratio is the ratio of the number of packets generated at the sources to the number of packets received by the destinations. This metric reflects the network throughput. One of our goals is to design an Energy-Efficient MAC Protocol which can improve energy consumption without suffering a significant loss of capacity. Thus, this metric is useful to measure any degradation in network throughput.

End-to-end delay: This metric includes not only the delays of data propagation and transfer, but also all possible delays caused by buffering, queuing, and retransmitting data packets.

Energy efficiency: We define energy efficiency as

Energy efficiency = Total number of bits transmitted / Total energy consumed

where the total bits transmitted is calculated using application-layer data packets only, and total energy consumption is the sum of each node's energy consumption during the simulation time. The unit of energy efficiency is bit/Joule, and the greater the number of bits per Joule, the better the energy efficiency achieved.

7.3 Transmission Power Management Schemes

Ad hoc is a kind of special wireless network mode. An ad hoc wireless network is a collection of two or more devices equipped with wireless communications and networking capability. Such devices can communicate with another device that is immediately within their radio range or one that is outside their radio range not relying on access point. A wireless ad hoc network is self-organizing, self-disciplining, and self-adaptive. The main characteristics of ad hoc networks are as follows:

Dynamic topology: Because nodes in the network can move arbitrarily, the topology of the network also changes.

The *bandwidth* of the link is unstrained, and the capacity of the network is also tremendously variable. Because of the dynamic topology, the output of each relay node will vary with the time, and then the link capacity will change with the link change. At the same time, compete-collision and interference make the actual bandwidth of ad hoc networks smaller than their bandwidth in theory.

Power limitation in mobile devices is a serious factor. Because of the mobility characteristics of the network, devices use batteries as their power supply. As a result, advanced power conservation techniques are very necessary in designing a system.

The safety is limited in a physical aspect. The mobile network is more easily attacked than the fixed network. Overcoming the weakness in safety and the new safety trouble in wireless networks are on demand.

7.3.1 Power Management of Ad Hoc Networks

The equipment in ad hoc networks always uses exhaustible energy as their power supply such as batteries. Despite the fact that mobile computing is evolving rapidly with advances in wireless communications and devices getting smaller and more efficient, advances in battery technology have not yet reached the stage where a mobile computer can operate for days without recharging. Therefore, advanced power conservation techniques are necessary. A variety of techniques can be used to cope with power scarcity. Table 7.1 lists some of these techniques at ad hoc networks' protocol layers. Based on the analysis of multicast routing in ad hoc networks, we propose a distributed multicast routing protocol—the Power Cost Calculate Balance (PCCB) Routing Protocol, which is based on the device's energy.

7.3.2 The Basic Idea of Power Cost Calculate Balance (PCCB) Routing Protocol

The PCCB Routing Protocol is a source-initiated, on-demand routing protocol. It aims to find out the minimum power-limitation route. The power limitation of a route is decided by the node which has the minimum energy in that route. So compared with the minimum node energy in any other route, the minimum node energy in the minimum power-limitation route has more energy. In other words, the value of that node's energy is the maximum of all minimum node energy in all selectable routes.

7.3.2.1 The Routing Process of PCCB Routing Protocol

7.3.2.1.1 Protocol Assumption

To simplify the model, the following assumptions are made:

1. A node *can* get the value of its current energy.
2. The links are bidirectional.

Table 7.1   Power Management at Various Protocol Layers

Protocol Layer	Power Conservation Techniques
Data-Link Layer	Avoid unnecessary retransmission. Avoid collision in channel access whenever possible.Put receive in standby mode whenever possible. Use or allocate contiguous slots for transmission and reception whenever possible. Turn radio off (sleep) when not transmitting or receiving.
Network Layer	Consider route relaying load. Consider battery life in route selection. Reduce frequency of sending control message. Optimize size of control headers. Efficient route reconfiguration techniques.
Transport Layer	Avoid repeated retransmissions. Handle packet loss in a localized manner. Use power-efficient error control schemes.
Application Layer	Adopt an adaptive mobile quality of service (QoS) framework. Move power-intensive computation from a mobile host to the base station. Use proxies for mobile clients. Proxies can be designed to make applications adapt to power or bandwidth constraints. Proxies can intelligently cache frequently used information, suppress video transmission and allow audio, and employ a variety of method to conserve power.

7.3.2.1.2   Route Discovery

Because the PCCB is a source-initiated on-demand routing protocol, nodes that are not on a selected path do not maintain routing information or participate in routing table exchanges.

The PCCB Routing Protocol uses the following fields with each route table entry:

- Destination node address
- Destination sequence number (guarantees the loop freedom of all routes toward that node)
- Valid destination sequence number *flag*
- Power boundary (the minimum energy of all nodes in the route)
- Hop count (the number of hops needed to reach a destination)
- Next hop
- Lifetime (the expiration or deletion time of the route)

The route discovery of the PCCB is as follows.

1. When the source node wants to send a message to the destination node and does not already have a valid route to that destination, it initiates a path discovery process to locate the other node. The source node disseminates a route

request (RREQ) to its neighbors. The RREQ includes such information as destination Internet Protocol (IP) address, destination sequence number, power boundary (the minimum energy of all nodes in the current found route), hop count, lifetime, and so forth. The destination sequence number field in the RREQ message is the last-known destination sequence number for this destination and is copied from the destination sequence number field in the routing table. If no sequence number is known, the unknown sequence number flag must be set. The power boundary is equal to the source's energy. The hop count field is set to zero. When the neighbor node receives the packet, it will forward the packet if it matches some conditions.

2. When a node receives the RREQ from its neighbors, it first increments the hop count value in the RREQ by one, to account for the new hop through the intermediate node if the packet should not be discarded. The originator sequence number contained in the RREQ must be compared to the corresponding destination sequence number in the route table entry. If the originator sequence number of the RREQ is not less than the existing value, the node compares the power boundary contained in the RREQ to its current energy to get the minimum, and then updates the power boundary of the RREQ with the minimum, which is the latest power boundary of this route.

   ■ If the originator sequence number contained in the RREQ is greater than the existing value in its route table, the relay node creates a new entry with the sequence number of the RREQ (the information can he acquired from the RREQ).

   ■ If the originator sequence number contained in the RREQ is equal to the existing value in its route table, the power boundary of the RREQ must be compared to the corresponding power boundary in the route table entry. If the power boundary contained in the RREQ is greater than the power boundary in the route table entry, the node updates the entry with the information contained in the RREQ (including power boundary and hop count).

   During the process of forwarding the RREQ, intermediate nodes record in their route tables the addresses of neighbors from which the first copy of the broadcast packet was received, thereby establishing a reserve path. If the same RREQs are later received, these packets are silently discarded. For each valid route maintained by a node as a routing table entry, the node also maintains a list of precursors that may be forwarding packets on this route. The list of precursors in a routing table entry contains those neighboring nodes to which a route reply was generated or forwarded.

3. Once the RREQ has arrived at the destination node or an intermediate node with an active route to the destination, the destination or intermediate node generates a route reply (RREP) packet and unicasts it back to the neighbor from which it received the RREQ. If the generating node is an intermediate node, it has an active route to the destination, the destination sequence number

in the node's existing route table entry for the destination is not less than the destination sequence number of the RREQ, and the "destination-only" flag is not set. If the generating node is the destination itself, it must update its own sequence number to the maximum of its current sequence number and the destination sequence number in the RREQ packet immediately before it originates an RREP in response to an RREQ. The destination node places its (perhaps newly incremented) sequence number into the destination sequence number field of the RREP and enters the value zero in the hop count field of the RREP. When generating an RREP message, a node wipes the destination IP address, originator sequence number, and power boundary from the RREQ message into the corresponding fields in the RREP message.

4. When a node receives the RREP from its neighbors, it first increments the hop count value in the RREP by one. As the RREP is forwarded back along the reverse path, the hop count field is incremented by one at each hop. Thus, when the RREP reaches the source, the hop count represents the distance, in hops, of the destination node from the source node. The originator sequence number contained in the RREP must be compared to the corresponding destination sequence number in the route table entry. If the originator sequence number of the RREP is not less than the existing value, the node compares the power boundary contained in the RREP to its current energy to get the minimum, and then updates the power boundary of the RREP with the minimum, which is the latest power boundary of this route.

   ■ If the sequence number in the routing table is marked as invalid in route table entry or the destination sequence number in the RREP is greater than the node's copy of the destination sequence number, the intermediate node creates a new entry with the destination sequence number of the RREP and marks the destination sequence number as valid. The power boundary field in the route table entry is set to the power boundary contained in the RREP.

   ■ If the originator sequence number contained in the RREP is equal to the existing destination sequence number in the node's route table, the power boundary of the RREP must be compared to the corresponding power boundary in the route table entry. If the power boundary contained in the RREP is greater than the node's copy of the power boundary, the power boundary in the entry is set to the value of the power boundary in the RREP.

The next hop in the route entry is assigned to be the node from which the RREP is received, which is indicated by the source IP address field in the IP header. The current node can subsequently use this route to forward data packets to the destination.

7.3.2.1.3 Route Maintenance

Inheriting from the Ad Hoc On-Demand Distance Vector (AODV) Routing Protocol, a node uses a Hello message, which is a periodic local broadcast by a node to inform each mobile node in its neighborhood to maintain the local connectivity. A node should only use Hello messages if it is part of an active route. If, within the past delete period, it has received a Hello message from a neighbor and then does not receive any packets from that neighbor (Hello messages or otherwise) for more than allowed-Hello-loss Hello-interval milliseconds, the node should assume that the link to this neighbor is currently lost. The node should send a route error (RERR) message to all precursors indicating which link is failed. Then the source initiates another route search process to find a new path to the destination or start the local repair.

7.3.3 Analysis of the PCCB Routing Protocol

The PPCB Routing Protocol is a pure on-demand routing protocol, as nodes that are not on a selected path do not maintain routing information or participate in routing table exchanges. PCCB allows mobile nodes to obtain routes quickly for new destinations and respond to link breakages and changes in network topology in a timely manner. The operation of PCCB is loop free and, by avoiding the "counting to infinity" problem, offers quick convergence when the ad hoc network topology changes (typically, when a node moves in the network). When links break, PCCB causes the affected set of nodes to be notified so that they are able to invalidate the routes using the lost link. As in the AODV, the shortest routing is found when the source initiates a route discovery with a new destination sequence number. But one distinguishing feature of PCCB is its use of a power boundary as a selection criterion. The power boundary is the minimum of all nodes' energy in the route. Using a power boundary ensures the updated route has the greater power boundary. Given the choice between two routes to a destination, a requesting node is required to select the one with the greatest power boundary.

The PCCB Routing Protocol selects the shortest path at first, which decreases the average relaying load for each node and therefore increases the lifetime of most nodes. At the same time, the PCCB Routing Protocol updates the route using the power boundary as metrics, which can prevent nodes from being unwisely overused by extending the time until the first node powers down and increasing the operation time before the network is partitioned. This avoids additional control overhead and power consumption to perform a new route discovery process to find a path to the destination.

When the energy is nearly exhausted, the Operating System (OS) and Basic Input–Output System (BIOS) will take actions in preparation for power down, which needs more power. So the maximum power boundary route can reduce the additional information operations and conserve energy. In a word, the PCCB Routing Protocol can optimize power utilization.

7.3.4 MAC Protocol

The primary task of the IEEE 802.11 MAC Protocol is to provide an efficient, fair, secure, and reliable data transfer service. Several problems arise if the communication medium is a radio channel. The MAC Protocol has to take the noisy and unreliable channel into account. This includes that the channel conditions may change rapidly. Stations may not be able to communicate directly, which leads to the hidden terminal problem and the exposed terminal problem. In the first case, communication is established, although it possibly interferes with a packet reception of neighboring stations. In the Exposed Terminal Scenario, no communication is established because of an ongoing communication between a pair of neighbor stations, although their communication would not be impaired. The IEEE 802.11 MAC Protocol deals with this and other problems, and fulfills the requirements appropriately as mentioned above. The basic MAC Protocol is similar to the 802.3 MAC Protocol.

7.3.5 Power Saving

Power consumption is crucial in mobile communications. Therefore, a wireless Network Interface Card (NIC) should consume as little power as possible. Power Saving (PS) is an optional and complex function, which can be used in Independent Basic Service Sets (IBSSs) or infrastructure BSSs (Basic Service Sets). The latter is not considered, because this thesis focuses on ad hoc networks. The principle of power saving in IEEE 802.11 is to turn off the transmission and reception hardware (NIC). Although IEEE 802.11 does not define under which circumstances a station can enter the PS mode, it is obvious to turn off the transmission and reception circuitry if no traffic has to be served. Unfortunately, several problems arise that can result in the opposite effect of power saving if not appropriately addressed. The "turn-off" mechanism creates two problems to be solved:

- How does a station send packets to another station in power-saving mode?
- How does a station in power-saving mode receive packets?

IEEE 802.11 power saving is based on buffering and synchronization. Packets destined to a station in PS mode have to be buffered until the station is "awake" to receive buffered packets. Stations have to be synchronized in such a manner that packets are only transmitted if the intended receiver is ready (awake) for reception. In infrastructure mode, buffering and synchronization are performed centrally by the Access Point (AP). It is more complicated in an IBSS network, because buffering

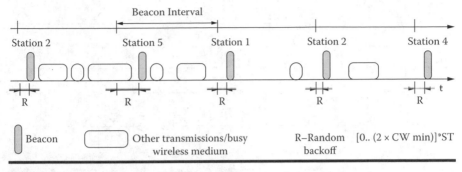

Figure 7.8   Beacon generation in an IBSS.

and synchronization have to be performed in a distributed manner. Eventually PS in IEEE 802.11 networks is implemented by means of a timing synchronization function and the actual PS function, which are described in the following subsections. Current wireless local area networks (WLANs) implement power saving for the ad hoc Distribution Coordination Function (DCF) mode.

7.3.6 Timing Synchronization Function

The purpose of the PS Timing Synchronization Function (TSF) is to provide globally known instants in an IBSS (or BSS) where stations in PS mode can be informed about pending traffic of other stations. Therefore, all stations within one IBSS have to be synchronized to a common clock. In an IBSS, timing synchronization is achieved in a distributed manner. All stations are capable to generate and send a beacon packet containing a copy of its own timer. Stations that receive a beacon adopt the time value if it is later than its own timing value. In the long run, all stations tend to have the "earliest" time of all nodes. Beacons have to be sent at a rate determined by the *BeaconPeriod*, which is contained in beacons and probe response messages. Stations joining an IBSS must adopt this value. The beacon interval is established by the station that started the IBSS. At the beginning of a beacon interval, all stations have to suspend ongoing backoff procedures of non-beacon traffic. All stations choose a random backoff delay from the interval [0 . . . (2 × CWmin)] × ST and back off for this time after the wireless medium becomes idle. If another beacon arrives before the backoff timer expires, the beacon transmission is canceled. If the backoff timer expires and no other beacon has arrived, the beacon is transmitted. Hence, convergence of the synchronization algorithm is only probabilistic. Figure 7.8 depicts the beacon generation process.

7.3.7 Power-Saving Function

The PS function describes the necessary steps for a station in PS mode to turn off the transmitter and receiver circuitry, to inform other stations or to retrieve

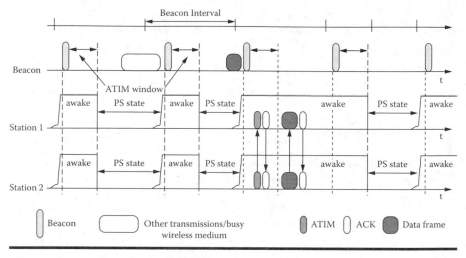

Figure 7.9 PS operation in an IBSS.

information about pending packets, and to transmit or receive traffic. The PS function is different for ad hoc and infrastructure networks. The PS function for an IBSS (ad hoc network) operates in a distributed manner. That is, frames destined to nodes in PS mode have to be buffered locally instead of at a centralized facility. A station that wants to enter the PS mode has to successfully complete a frame exchange with another station with the power bit set in the frame header. Note that neither a specific station nor all stations need to be informed. Once the frame exchange has been successfully completed, the station may enter the PS state. In the PS state, the station has to wake up periodically at the estimated time of a beacon transmission. The station is further required to stay "awake" for a period referred to as the ATIM window to receive announcements from other stations with buffered traffic. The ATIM window can be used by the station itself to announce buffered traffic. If there is no traffic announcement for the station, the PS state is reentered. If an announcement in the form of an ATIM arrives, the station has to acknowledge the ATIM and has to stay "awake" beyond the ATIM window until all of the announced buffered frames were received. During the data frame transmission period that follows the ATIM window, announced traffic can be sent following the basic channel access rules. Both form the beacon interval. The PS specification requires a station desiring to transmit a frame to another station to estimate the power-saving status of the other station. The estimate can be based on local information or the last data frame exchange with that station. The IEEE standard leaves open the solution to the problems of how the estimate is created and on which information the estimate is based.

Figure 7.9 shows two stations operating in PS mode. Both stations are required to wake up upon every beacon and to listen throughout the ATIM window. Station 2 wishes to send a frame to station 1 and therefore announces that frame by an

ATIM. Station 1 responds with an ACK, whereupon station 2 starts data frame transmission after completion of the ATIM window, the PS state, the sending station creates an ATIM, which is sent during the ATIM window. Once the ACK of the intended station is received, the station sends the data frame after completion of the ATIM window. It is assumed that a station always is in PS mode if it has not explicitly indicated that it will stay awake by an ACK in response to an ATIM (even in the case that it is awake because it has pending traffic). Multicast frames must also be announced by the sending station during the ATIM window before transmitting the data frame. Both multicast ATIM and multicast data frame do not need to be acknowledged. The transmission of ATIMs during the ATIM window is randomized by means of the backoff procedure as described before. The backoff procedure uses a Congestion Window (CW) size of exactly CWmin independent of whether it is an initial transmission of an ATIM or a retry. The transmission of data frames follows the normal DCF access procedure.

7.3.8 *Power-Saving Potential*

There are significant differences in the PS function for infrastructure BSSs and IBSSs, which lead to differences in their PS potential. PS in the IBSS places a larger burden on the mobile station than in the BSS with AP case, because (mobile) stations are responsible for beacon management and transmission. In addition, announcements (ATIMs) have to be generated and transmitted, and corresponding ACKs have to be received to transmit data frames. The buffering of frames also has to be performed by every station in PS mode. In a BSS, these functions are performed centrally by the APs, which normally do not suffer from energy limitation because of a wired power supply. Another significant difference is the ratio of PS state and "awake" state holding times, which tends to be more advantageous for BSS with AP. A station operating in a BSS with AP does not need to wake up every beacon time. Furthermore, stations can return to the PS state immediately after the beacon instead of staying awake for an ATIM window to listen for announcements, because the AP sends the announcements for buffered packets within the beacon without requiring acknowledgments. The protocol overhead incurred by the backoff procedure prior to sending the announcements as well as possible collisions limit the PS potential. This overhead does not exist in a BSS with AP because generation of beacons and announcements is centrally managed by the AP. Another reason is the vulnerability of wireless transmissions regarding the transmission range. It can be assumed that the AP has an exposed placement within an infrastructure BSS, whereas IBSS stations may suffer from range problems. This leads to necessary repetitions of ATIMs and data frames and less time to be in the PS state. Note that PS in an IBSS does not cover the case of hidden terminals. In this case, PS cannot be applied because of synchronization inconsistencies within the network.

7.4 Transmission Power Control

In the majority of cases, transmission power control is used to optimally communicate with a given receiver, for example achieving a certain signal-to-noise ratio (SNR) under the constraint of a low interference to other communicating nodes. The signal amplification process is a power-intensive process. A very simple method for saving energy is the optimization of the amplifier to efficiently operate at the most commonly encountered transmission power. But good amplifier efficiency on its own does not necessarily result in low power consumption, and the radio frequency (RF) power has to be adjusted to cope with changing radio channel characteristics. Depending on the RF hardware, dynamically choosing transmission power can also save energy. In traditional cellular environments, transmission power is adapted to account for signal degradation and interference (e.g., in Code Division Multiple Access [CDMA] systems). However, energy efficiency is usually not regarded as the primary optimization goal. In ad hoc networks, the situation is usually somewhat different.

Ad hoc networking is often identified with multihop communication, and therefore power assignments are not necessarily controlled by a single (central) entity. If centralized knowledge is available, the necessary power setting is determined as described in Section 7.3.7, simultaneously controlling the network topology. Other approaches attempt to only use local information, either by exploiting them heuristically or by integrating some additional data such as location or direction to neighboring nodes to find optimum transmission power values by using distributed algorithms. Of high interest from a reliable and scalable point of view are the distributed algorithms of Rodoplu and Meng, who try to find the most energy-efficient neighbors to be used as relays1 in the vicinity of a node, and Wattenhofer, who also tries to find the most energy-efficient neighbors using the angle of the received signal to estimate direction. These algorithms face a number of open problems: how are the time-varying characteristics of the radio channel taken into account by these algorithms? What about mobility? How does additional information like location correlate with the radio channel? What is the performance penalty if estimations are incorrect?

7.4.1 Adapting Transmission Power to the Channel State

Depending on the channel state, transmission power can also be adapted on short time scales. Thereby, if the channel quality is estimated as better, a lower transmission power is assigned, resulting in less interference and power consumption by the amplifier. On the other hand, if the channel quality is estimated as worse, a higher transmission power is assigned or the transmission process might be stopped temporarily until the communication channel is good again.

7.4.2 MAC Techniques

Apart from deciding how much transmission power to use, the decision whether power should be invested at all is crucial, for example after a transmission fails due to collisions or bad channel conditions. The MAC layer decides about the point in time when a transmission starts.

> *On and off switching of network interface and processor*: A promising method to save energy is to shut down nodes or network interfaces within a wireless network. The decision of when to shut down can depend on the network as a cooperative system, or local information where each node optimizes the energy consumption for itself. The main problem is uncertainty of when the network interface has to be turned on again in combination with the fact that powering up an interface incurs both time and energy overhead. Power-saving protocols as applied in common IEEE 802.11 WLAN cards decide to turn off the network interface based on the traffic need. If a node has to transmit or receive data, it will stay awake; otherwise, it can go asleep for a while to save energy. A node periodically has to check whether packets have to be sent or received, which requires a centralized or more complex distributed synchronization algorithm. In a centrally scheduled communication system, for example HIPERLAN, some of these decisions are simplified. Because of the addition a connectivity and adapts its participation in the multihop network according to the measured operating region with the objective to minimize the network energy consumption. Similar concepts can be applied to the processor or other components of a mobile device or network interface. The processor is a particularly promising component, which can additionally enable a gradual adaptation of energy consumption and delivered performance by techniques like dynamic voltage scaling. Such gradual strategies can be more effective than completely turning off a processor. However, a certain amount of previous knowledge about processing requirements is necessary. This information is available for many types of interactive applications. As many processor cores for embedded devices support this concept (e.g., Intel StrongArm), dynamic voltage scaling is promising for practical applications and could also be used for MAC and base-band processors.
>
> *Avoiding collision*: Collisions may have several reasons. One reason might be the protocol itself, which is explicitly designed to deal with (rare) collisions. Another reason is the network topology, which may lead to hidden terminals. Collisions in turn cause retransmissions and waste of energy. Hidden terminals can be avoided by using a derivate of the Power-Aware Medium Access Control with Signaling (PAMAS) Protocol, which is a more complex derivate of the Multiple Access Collision Avoidance (MACA) Protocol. It belongs to the class of single-channel-based busy tone protocols. MAC Protocols, which are explicitly designed to work in an energy-efficient manner, will avoid or

minimize the likelihood of collisions during data transfer. An example is the Energy-Conserving MAC (EC-MAC) Protocol. Centralized MAC Protocols like HIPERLAN/2 also belong to this category, as do other time division multiple access (TDMA)–based systems. The downside of these protocols is the additional synchronization overhead, which requires some energy as well. The trade-off between these two aspects is crucial in evaluating energy efficiency.

Rate adoption: Another MAC option to conserve energy is to adapt the transmission rate according to varying channel conditions for the sake of an optimized throughput and higher energy efficiency. If the channel quality decreases, a more robust modulation scheme is used which usually results in lower transmission rates. Rate adaptation may also be used to avoid energy-demanding radio processing. For instance, in HIPERLAN packet headers are sent at 1 mbps while the payload information is sent at a rate of 1 overhead in both time and energy to turn a transmitter on again; the scheduling periods (e.g., two milliseconds for HIPERLAN) can be too short to justify an aggressive turning off. Turning off network interfaces can also be applied to multihop networks. For instance, the ASCENT Protocol is a distributed protocol where each node assesses its 23 mbps; IEEE 802.11 follows a similar paradigm. Processing received data with high data rates consumes considerably more energy than with low data rates. Therefore, based on the control information of the packet header sent at 1 mbps (or 2 mbps), a node can decide whether the payload should actually be received.

7.4.3 Logical Link Control

Whereas the MAC layer decides when to attempt to send a packet, the link-layer control (LLC) determines the exact form of the packet, for example its size and how to deal with errors. Most of the link-layer procedures have a direct impact on the energy efficiency of a communication system.

Packet size adaptation: On the link-layer protocol level, the packet size may be adapted according to the channel characteristics as proposed by Modiano. Modiano developed an ARQ Protocol that automatically adapts the packet size to changing channel conditions (bit error rates, or BERs) with the objective to improve the performance. As a side effect, energy consumption is decreased, whereas performance is increased. The challenge for this protocol is to obtain a sufficiently good estimate of the channel condition and the predictive power of such estimations.

Combined FEC and ARQ schemes: Whereas Automatic Repeat Request (ARQ) schemes simply repeat packets in the case of transmission errors, Forward Error Correction (FEC) contributes to a more robust transmission of data with an overhead penalty. There is interval trade-off between these two techniques in terms of energy conservation, which varies with changing the channel conditions. Therefore, Lettieri and Srivastava proposed a combined FEC–ARQ scheme. The trade-off between the number of retransmissions and the longer packets with FEC is evaluated, and an algorithm is proposed that adapts both values for energy efficiency.

Packet pacing: An adaptive energy-efficient ARQ Protocol that slows down the transmission rate (pace of packet transmission) when the channel is impaired and vice versa is proposed by Zorzi. Additionally, he proposes a kind of channel probing where the channel is tested with short low-power packets at a certain pace as long as the channel is impaired and data transmission is resumed at high power when the channel is assumed to be good again. This is exemplary of a basic approach: do not waste energy on a bad channel, as attempting to overcome the bad channel is prohibitively expensive.

Asymmetric protocols: Another approach toward energy-efficiency and lifetime extension of mobile nodes is the asymmetric protocol design. An example for this approach is the AIRMAIL Protocol. Protocol processing and scheduling in conjunction with a combination of FEC and ARQ are mainly performed at a central facility (base station) having extended power sources or access to a fixed power supply. This concept is commonly used by most cellular systems, and in a sense it is extended to multihop communication systems where responsibilities are exchanged or carried forward among nodes, for example depending on the currently available resources. Responsibilities are often assigned to so-called clusterheads, which temporarily take over certain tasks in the context of relaying and routing.

Exploiting channel predictions: If the future channel state can be predicted with some minimal accuracy even for a rather short period of time, this knowledge will evidently be exploited to improve the protocol's energy efficiency by postponing transmissions, modifying transmission power, selecting appropriate modulation schemes, and so forth. The main questions to answer here are as follows:

- Does the prediction process result in any additional power consumption on its own? For example, does prediction involve active probing?
- What is the accuracy of the prediction process?
- What is the consequence of incorrect predictions of the energy efficiency?

This is currently a field of active research, and only few actual results are known in this field.

Pulsed battery discharge: From the battery point of view, a pulsed discharge of the battery can considerably extend the operation time of a wireless node. Battery relaxation phases may be achieved by on and off switching of the interface hardware (see above), and can be sufficiently triggered by packet pacing or packet scheduling. The premise is a separate battery for the network interface or an intelligent battery control that appropriately controls the discharge of single battery cells, which in turn are fixed or dynamically assigned to certain hardware parts of a mobile node.

Routing: In traditional cellular systems, the routing problem changes to the handover problem: Which access point should be selected to serve a mobile device? As a handover decision is usually based on the channel quality between mobile devices and access points, energy efficiency is at least implicitly considered in this process. From a system perspective, it could be conceivable that a handover decision that is energy conserving for a particular mobile device could be suboptimal when considering all mobiles together; this suboptimality could, for example, be due to a different interference situation in neighboring cells. The routing problem becomes much more difficult if multihop radio communication is considered. In such a multihop system, it is no longer clear which sequence of nodes should be traversed to reach a given destination. Several optimization metrics can be introduced to support this choice. A number of routing protocols have been developed to meet the specific needs of such multihop networks, for example proactive protocols like Destination-Sequenced Distance Vector (DSDV), which periodically sends route updates to learn all routes to a destination in the network; reactive protocols like Dynamic Source Routing (DSR) and Power-Aware Routing Optimization (PARO), which start searching for the destination only if there is a packet to transmit; and hybrid schemes like AODV, FSR, and Temporally Ordered Routing Algorithm (TORA), but energy efficiency is not the prime target of these protocols. More recently, energy efficiency has moved into focus, particularly motivated by the vision of wireless sensor networks. A frequently used concept is to assign routing and forwarding responsibilities to a node acting on behalf of a group of nodes (a "cluster"); routing and forwarding then only take place among these "clusterheads." The choice of clusterheads can be based on the availability of resources (battery capacity) and is rotated among several nodes in many approaches. Examples of such clustered protocols are the Zone Routing Protocol (ZRP) and Low-Energy Adaptive Clustering Hierarchy (LEACH). Additionally, some routing protocols take the physical location of nodes into account (e.g., Geographical Adaptive Fidelity, or GAF). The challenge for all of these multihop routing protocols is the evaluation of the trade-off between energy savings by clever routing and the overhead required to obtain the routing information, particularly in the face of uncertainties induced by mobility, time-varying channels, and so forth.

Packet size–dependent energy-efficient power control: One of the most power-demanding components of a WLAN NIC is the RF power amplifier. The RF power amplifier can consume more than 50 percent of the overall power consumption of the network interface. The power consumption caused by the amplifier is relatively high, but it is controllable within a reasonable range.

Two basic energy-saving strategies follow from the facts above. First, the RF power amplifier should be turned off when not in use. This feature is realized in any implementation of a WLAN card. The entire transmit circuitry including the RF power amplifier, the RF–Intermediate Frequency (IF) converter, and the IF modem is turned off while the WLAN NIC is not sending. Additionally, almost all parts except the WLAN NIC are put into sleep mode or are turned off if power saving is supported and no data is to be transmitted or received. Second, the RF transmission power might be reduced to decrease the power consumption of the RF power amplifier, although this may indirectly increase the overall power consumption. The authors believe that there is a large potential for energy saving if the RF transmission power level is properly controlled according to the packet size.

The basic idea is as follows. The reduction of the RF output power results in a reduction of the instantaneous power consumption of the WLAN NIC and the RF power amplifier in particular. This is certainly traded off to a certain extent by bit errors. An increase of the BER unavoidably leads to a higher probability of retransmissions on the MAC, data-link, or transport layer, which in turn increases the energy consumption. On the other hand, an irreflective high RF power level setting could also lead to needless power consumption, because the BER does not improve significantly after a certain point. Therefore the RF output power adjustment should be balanced with the number of retransmissions to achieve the lowest energy consumption. This requires a power control mechanism. It is shown for a dedicated network scenario that energy savings during the transmission process can be achieved using a simple power control algorithm. The algorithm employs a selection of the appropriate transmission power level according to the size of the next MAC frame to be transmitted.

7.5 Ad Hoc On-Demand Distance Vector (AODV) Protocol

AODV Routing Protocol is a reactive routing algorithm. It maintains the established routes as long as they are needed by the sources. AODV uses sequence numbers to ensure the freshness of routes. Route discovery and route maintenance for AODV are described below.

7.5.1 Route Discovery

The route discovery process is initiated whenever a traffic source needs a route to a destination. Route discovery typically involves a networkwide flood of Route Request (RREQ) packets targeting the destination and waiting for a Route Reply (RREP). An intermediate node receiving an RREQ packet first sets up a reverse path to the source using the previous hop of the RREQ as the next hop on the reverse path. If a valid route to the destination is available, then the intermediate node generates an RREP; otherwise, the RREQ is rebroadcast. Duplicate copies of the RREQ packet received at any node are discarded. When the destination receives an RREQ, it also generates an RREP. The RREP is routed back to the source via the reverse path. As the RREP proceeds toward the source, a forward path to the destination is established.

7.5.2 Route Maintenance

Route maintenance is done using route error (RERR) packets. When a link failure is detected, an RERR is sent back via separately maintained predecessor links to all sources using that failed link. Routes are erased by the RERR along its way. When a traffic source receives an RERR, it initiates a new route discovery if the route is still needed. Unused routes in the routing table are expired using a timer-based technique.

7.6 Local Energy-Aware Routing Based on AODV (LEAR-AODV)

The first on-demand routing protocol proposed is called Local Energy-Aware Routing Based on AODV (LEAR-AODV). The main objective is to balance energy consumption among all participating nodes. A similar mechanism is used in extending Dynamic Source Routing (DSR) protocol. In their approach, each mobile node relies on local information about the remaining battery level to decide whether to participate in the selection process of a routing path or not. An energy-hungry node can conserve its battery power by not forwarding data packets on behalf of others. The decision-making process in LEAR-AODV is distributed to all relevant nodes. Route discovery and route maintenance for LEAR-AODV are described below.

7.6.1 Route Discovery

In AODV, each mobile node has no choice and must forward packets for other nodes. In LEAR-AODV, each node determines whether or not to accept and forward the RREQ message depending on its remaining battery power (Er). When it is lower than a threshold value θ ($Er \le \theta$), the RREQ is dropped; otherwise, the message is forwarded. The destination will receive a route request message only when all intermediate nodes along the route have enough battery levels.

7.6.2 Route Maintenance

Route maintenance is needed either when the connections between some nodes on the path are lost due to node mobility, or when the energy resources of some nodes on the path are depleting too quickly. In the first case, and as in AODV, a new RREQ is sent out, and the entry in the route table corresponding to the node that has moved out of range is purged. In the second case, the node sends an RERR back to the source even when the condition $Er \leq \theta$ is satisfied. This route error message forces the source to initiate route discovery again. This is a local decision because it is dependent only on the remaining battery capacity of the current node. However, if this decision is made for every possible route, the source will not receive an RREP message even if a route exists between the source and the destination. To avoid this situation, the source will resend another RREQ message with an increased sequence number. When an intermediate node receives this new request, it lowers its θ by d to allow the packet forwarding to continue. We use a new control message, *ADJUST_Thr*. When a node drops an RREQ message, it instead broadcasts a *ADJUST_Thr* message. The subsequent nodes closer to the destination now know that a request message was dropped and lower their threshold values. Now, the second route request message can reach the destination. When the destination receives an RREQ, it generates an RREP. As in AODV, the RREP is routed back to the source via the reverse path. We notice that LEAR-AODV interworks easily with AODV. By this, we mean that an ad hoc network can contain both nodes carrying out LEAR-AODV and nodes carrying out AODV as routing protocols.

7.7 Power-Aware Routing Based on AODV (PAR-AODV)

The second on-demand routing protocol we propose is called Power-Aware Routing Based on AODV (PAR-AODV). The main objective is to extend the useful service life of an ad hoc network. PAR-AODV solves the problem of finding a route Π, at route discovery time t, such that the following cost function is minimized:

$$C(\Pi, t) = \Sigma \, C_i \, (\, t \,) \qquad (7.1)$$

where

$$C_i \, (\, t \,) = \rho_t \, [F_t \, / \, E_t \, (\, t \,)]^\alpha \qquad (7.2)$$

and where

ρ_i is the transmit power of node i,
F_i is the full-charge battery capacity of node i,
$E_i(t)$ is the remaining battery capacity of node i at time t, and
α is a positive weighting factor.

The route discovery and route maintenance for PAR-AODV are described below.

7.7.1 Route Discovery

In PAR-AODV, activity begins with the source node flooding the network with RREQ packets when it has data to send. All nodes except the source and the destination calculate their link cost, Ci, using Equation 7.2 and add it to the path cost in the header of the RREQ packet (Equation 7.1). When the destination node receives an RREQ packet, it sends an RREP packet to the source. When an intermediate node receives an RREQ packet, it keeps the cost in the header of that packet as *Min-Cost*. If additional RREQs arrive with the same destination and sequence number, the cost of the newly arrived RREQ packet is compared to the *Min-Cost*:

- If the new packet has a lower cost and if the intermediate node does not know any valid route to the destination, *Min-Cost* is changed to this new value and the new RREQ packet is rebroadcast.
- If the new packet has a lower cost but the intermediate node knows a route to the destination, the node forwards (unicasts) a *COMPUTE_Cost* message. The *COMPUTE_Cost* calculates this route cost.
- Otherwise, if the new packet has a greater cost, the new RREQ packet is dropped.
- When the destination receives either an RREQ or a *COMPUTE_Cost* message, it generates an RREP message. The RREP is routed back to the source via the reverse path. This reply message contains the cost of the selected path. The source node will select the route with the minimum cost.

7.7.2 Route Maintenance

The route maintenance in PAR-AODV is the same as in LEAR-AODV. Hence, in PAR-AODV, when any intermediate node has a lower battery level than its threshold value ($Er \le \theta$), any request is simply dropped.

7.8 Lifetime Prediction Routing Based on AODV (LPR-AODV)

The last on-demand routing protocol we propose is called Lifetime Prediction Routing Based on AODV (LPR-AODV). This protocol favors the route with maximum lifetime, that is, the route that does not contain nodes with a weak predicted lifetime. LPR-AODV solves the problem of finding a route ρ at route discovery time t, such that the following cost function is maximized:

$$Max_{\pi}\left(T_{\pi}\left(t\right)\right)= Max_{\pi}\left(Min_{\pi}\left(T_{\pi}\left(t\right)\right)\right) \tag{7.3}$$

where $T_{\pi}(t)$ is the lifetime of path π, and $T_{\pi}(t)$ is the predicted lifetime of node i in path π.

LPR-AODV uses battery lifetime prediction. Each node tries to estimate its battery lifetime based on its past activity. This is achieved using a recent history of node activity. When node i sends a data packet, it keeps track of the residual energy value ($Ei(t)$) and the corresponding time instance (t). This information is recorded and stored in the node. After N packets are sent or forwarded, node i gets the time instance when the Nth packet is sent or forwarded (t') and the corresponding residual energy value ($Ei(t')$). This recent history, {(t, $Ei(t)$), (t', $Ei(t')$)}, is a good indicator of the traffic crossing the node. Hence, we use it for lifetime prediction. Our approach is a dynamic distributed load balancing approach that avoids power-congested nodes and chooses paths that are lightly loaded. Route discovery and route maintenance for LPR-AODV are described below.

7.8.1 Route Discovery

In LPR-AODV, all nodes except the destination and the source calculate their predicted lifetime, Ti. In each request, there is another field representing the minimum lifetime (*Min-lifetime*) of the route. A node i in the route replaces the *Min-lifetime* in the header with Ti if Ti is lower than the existing *Min-lifetime* value in the header.

$$T_i(t)=\frac{E_i(t)}{disch\arg e\_rate_i(t)} \tag{7.4}$$

where

discharge_rate$_i$(t) = $\dfrac{E_i(t^i)-E_i(t)}{t-t'}$ and $E_i(t)$ is the remaining energy of node i at time t;

t: current time corresponding to the moment when the node I sends or forwards the current packet; and

t': the recorded time instance corresponding to the moment when the Nth "predecessor" to the current packet was sent or forwarded by node i.

More precisely, when an intermediate node receives the first RREQ packet, it keeps the *Min-lifetime* in the header of that packet as *Min-lifetime*. If additional RREQs arrive with the same destination and sequence number, the *Min-lifetime* of the newly arrived RREQ packet is compared to the *Min-lifetime*:

- If the new packet has a greater *Min-lifetime* and if the intermediate node does not know any valid route to the destination, *Min-lifetime* is changed to this new value and the new RREQ packet is rebroadcast.
- If the new packet has a greater *Min-lifetime* but the intermediate node knows a route to the destination, the node forwards (unicasts) a *COMPUTE_lifetime* message. The *COMPUTE_lifetime* calculates this route lifetime.
- Otherwise, if the new packet has a lower *Min-lifetime*, the new RREQ packet is dropped.
- When the destination receives either an RREQ or a *COMPUTE_lifetime* message, it generates an RREP message. The RREP is routed back to the source via the reverse path. This reply message contains the lifetime of the selected path. The source node will select the route with the maximum lifetime.

7.8.2 Route Maintenance

As in the first algorithms, route maintenance is needed either when a node becomes out of direct range of a sending node or when there is a change in its predicted lifetime. In the first case (node mobility), the mechanism is the same as in AODV. In the second case, the node sends an RERR back to the source even when the predicted lifetime goes below a threshold level δ ($Ti(t) = \delta$). This route error message forces the source to initiate route discovery again. This decision depends only on the remaining battery capacity of the current node and its discharge rate. Hence, it is a local decision.

However, the same problem as in LEAR-AODV can occur. If the condition $Ti(t) = \delta$ is satisfied for all the nodes, the source will not receive a single reply message even though there exists a path between the source and the destination. To prevent this, we use the same mechanisms used in LEAR-AODV described above.

7.9 Summary

In this chapter, how to manage energy efficiently in wireless ad hoc networks is discussed. Because the nodes in MANETs are mobile and can be used for emergency purposes like military or natural disasters, each node should utilize its battery efficiently. Some of the problems which are faced while managing energy at the nodes in MANETs are limited energy reserve, difficulties in replacing batteries, lack of central coordination, and constraints on the battery source. This chapter deals with the issues concerning energy management. The challenge is not to provide each node with higher battery power but to utilize the available battery power in a very efficient manner.

This chapter deals with the different energy-efficient routing protocols which try to meet the challenge of using battery power efficiently. This chapter mainly deals with battery power management schemes and transmission power management.

7.10 Problems

7.1 Why is energy management needed in ad hoc networks?

7.2 Discuss the main reasons for energy management in ad hoc networks.

7.3 How is classification done in energy management schemes for ad hoc networks?

7.4 Explain the battery management schemes for ad hoc networks.

7.5 Give an overview of battery technologies used for ad hoc networks.

7.6 Explain the principles of battery discharge.

7.7 Discuss the impact of discharge characteristics on battery capacity.

7.8 Explain how battery modeling captures the characteristics of real-life batteries.

7.9 Discuss various battery models that are used.

7.10 Describe battery-efficient system architectures in detail.

7.11 Explain the typical implementation of a smart battery system with a simple example.

7.12 Discuss the Energy-Efficient Routing Protocol used in mobile ad hoc networks.

7.13 Give the overview of the IEEE 802.11 Power-Saving Mode.

7.14 Explain the EE-MAC Protocol with a suitable illustration.

7.15 Discuss the design criteria and features of the EE-MAC Protocol.

7.16 List power management at various protocol layers.

7.17 Explain the basic idea of the PCCB Routing Protocol.

7.18 Give an analysis of the PCCB Routing Protocol.

7.19 Explain the timing synchronization function and power-saving function used in power-saving mechanisms.

7.20 Describe the Power-Aware Routing Based on AODV (PAR-AODV) Protocol.

7.21 Discuss the Lifetime Prediction Routing Based on AODV (LPR-AODV) Protocol.

Bibliography

S. Banerjee, A. Misra, Minimum energy paths for reliable communication in multi-hop wireless networks, Proceedings of Annual Workshop on Mobile Ad Hoc Networking and Computing (MobiHOC 2002), 2002, pp. 146–156.

J-H. Chang, L. Tassiulas, Energy conserving routing in wireless ad-hoc networks, Proceedings of the Conference on Computer Communications (IEEE Infocom 2000) 2000, pp. 22–31.

B. Chen, K. Jamieson, R. Morris, H. Balakrishnan, Span: an energy-efficient coordination algorithm for topology maintenance in ad hoc wireless networks, Proceedings of International Conference on Mobile Computing and Networking (MobiCom'2001), 2001, pp. 85–96.

S. Doshi, S. Bhandare, T. X. Brown, An on-demand minimum energy routing protocol for a wireless ad hoc network, *Mobile Computing and Communications Review*, 6(2):50–66, 2002.

A. Ephremides, Energy concerns in wireless networks. *IEEE Wireless Communications*, 9(4):48–59, 2002.

G. Forman, J. Zahorjan, The challenges of mobile computing, *IEEE Computer*, 27(4):38–47, 1994.

G. Girling, J. Wa, P. Osborn, R. Stefanova, The design and implementation of a low power ad hoc protocol stack. *IEEE Personal Communications*, 4(5):8–15, 1997.

A. J. Goldsmith, S. B. Wicker, Design challenges for energy constrained ad hoc wireless networks, *IEEE Wireless Communications*, 9(4): 8–27, 2002.

D. Johnson, D. Maltz, Dynamic source routing in ad hoc wireless networks, in *Mobile Computing*, T. Imielinski, H. Korth (eds.), Kluwer Academic: 1996, pp. 153–181.

C. E. Jones, K. M. Sivalingam, P. Agrawal, J. C. Chen, A survey of energy efficient network protocols for wireless networks, *Wireless Networks*, 7(4):343–358, 2001.

J. Jubin, J. Tornow, The DARPA packet radio network protocols, *Proceedings of the IEEE* 75(1):21–32, 1987.

A. Kamerman, L. Monteban, WaveLAN-II: A high-performance wireless LAN for the unlicensed band, *Bell Labs Technical Journal*, 1997, pp. 118–133.

Q. Li, J. Aslam D. Rus, Online power-aware routing in wireless ad-hoc networks, Proceedings of International Conference on Mobile Computing and Networking (MobiCom'2001) 2001, pp. 97–107.

S. Narayanaswamy, V. Kawadia, R. S. Sreenivas, P. R. Kumar, Power control in ad-hoc networks: theory, architecture, algorithm and implementation of the COMPOW protocol, Proceedings of European Wireless 2002, pp. 156–162.

G. Pei, M. Gerla, T-W. Chen, Fisheye state routing: a routing scheme for ad hoc wireless networks, Proceedings of IEEE International Conference on Communications (ICC) 2000, pp. 70–74.

C. Perkins, *Ad Hoc Networking*, Addison-Wesley: Reading, MA, 2001, 1–28.

C. Perkins, P. Bhagwat, Highly dynamic destination sequenced distance-vector routing (DSDV) for mobile computers, *Computer Communications Review*, 24(4):234–244, 1994.

C. Perkins, E. Royer, Ad-hoc on-demand distance vector routing, Proceedings of 2nd IEEE Workshop on Mobile Computing Systems and Applications. 1999, pp. 90–100.

R. Ramanathan, R. Rosales-Hain, Topology control of multihop wireless networks using transmit power adjustment. Proceedings of the Conference on Computer Communications (IEEE Infocom 2000), 2000, pp. 404–413.

V. Rudolph, T. H. Meng, Minimum energy mobile wireless networks, *IEEE Journal of Selected Areas in Communications*, 17(8):1333–1344, 1999.

M. Sanchez, P. Manzoni, Z. H. Haas, Determination of critical transmission range in ad-hoc networks, Proceedings of Multiaccess, Mobility and Teletraffic for Wireless Communications (MMT'99), 1999.

J. Schiller, *Mobile Communications*, Addison-Wesley: Reading, MA, 2000, 161–214.

S. Singh, M. Woo, C. Raghavendra, Power-aware routing in mobile ad hoc networks, Proceedings of International Conference on Mobile Computing and Networking (MobiCom'98) 1998, pp. 181–190.

I. Stojmenovic, X. Lin, Power-aware localized routing in wireless networks, *IEEE Transactions on Parallel and Distributed Systems*, 12(11):1122–1133, 2001.

C-K. Toh, Maximum battery life routing to support ubiquitous mobile computing in wireless ad hoc networks, *IEEE Communications*, 39(6):138–147, 2001.

R. Wattenhofer, L. Li, P. Bahl, Y-M. Wang, Distributed topology control for power efficient operation in multihop wireless ad hoc networks, Proceedings of the Conference on Computer Communications (IEEE Infocom 2001), 2001, pp. 1388–1397.

H. Woesner, J-P. Ebert, M. Schlager, A. Wolisz, Power-saving mechanisms in emerging standards for wireless LANs: the MAC level perspective, *IEEE Personal Communications*, 5(3):40–48, 1998.

K. Woo, C. Yu, H. Y. Youn, B. Lee, Non-blocking, localized routing algorithm for balanced energy consumption in mobile ad hoc networks, Proceedings of International Symposium on Modeling, Analysis and Simulation of Computer and Telecommunication Systems (MASCOTS 2001) 2001, pp. 117–124.

Y. Xu, J. Heidemann, D. Estrin, Geography-informed energy conservation for ad hoc routing, Proceedings of International Conference on Mobile Computing and Networking (MobiCom'2001), 2001, pp. 70–84.

Chapter 8

Mobility Models for Multihop Wireless Networks

8.1 Introduction

In the performance evaluation of a protocol for an ad hoc network, the protocol should be tested under realistic conditions including, but not limited to, a sensible transmission range, limited buffer space for the storage of messages, representative data traffic models, and realistic movements of the mobile users (i.e., a mobility model). In this chapter, several mobility models are described that represent mobile nodes whose movements are independent of each other (i.e., Entity Mobility Models), and several mobility models are described that represent mobile nodes whose movements are dependent on each other (i.e., Group Mobility Models). The goal of this chapter is to present a number of mobility models to offer researchers more informed choices when they are deciding upon a mobility model to use in their performance evaluations, and present simulation results that illustrate the importance of choosing a mobility model in the simulation of an ad hoc network protocol. Specifically, how the performance results of an ad hoc network protocol drastically change as a result of changing the mobility model simulated is described.

A mobile ad hoc network (MANET) represents a system of wireless mobile nodes that can freely and dynamically self-organize into arbitrary and temporary network topologies, allowing people and devices to seamlessly communicate without any preexisting communication architecture. An ad hoc routing protocol is a convention or standard that controls how nodes come to agree on which way to route packets between computing devices in a MANET. Nodes do not have a priori knowledge of the topology of the network around them; they have to discover it. The basic idea is that a new node announces its presence and listens to broadcast announcements from its neighbors. The node learns about new near nodes and ways to reach them, and announces that it can also reach those nodes. As time goes on, each node knows about all the other nodes and one or more ways of how to reach them. We show from our simulation results that mobility models have a considerable effect on the performance of these routing protocols. Ad hoc networks are formed spontaneously and deployed during an emergency. The mobile nodes in the networks will freely move and communicate with each other. Due to high mobility and dynamic changing topology, we need to use a suitable mobility model for a particular network scenario. The change in mobility model affects the performance of routing protocols.

8.2 Mobility Models

A mobility model should attempt to mimic the movements of real mobile nodes. Changes in speed and direction must occur, and they must occur in reasonable time slots. For example, we would not want mobile nodes to travel in straight lines at constant speeds, because real mobile nodes would not travel in such a restricted manner. In this section, different types of mobility models are described.

Mobility models mainly are of two types:

1. Entity Mobility Model
2. Group Mobility Model

In this section, we present entity mobility models.

8.2.1 Entity Mobility Model

Entity Mobility Models represent mobile nodes whose movements are independent of each other. Examples of Entity Mobility Models are as follows:

Random Walk Mobility Model: A simple mobility model based on random directions and speeds

Random Waypoint Mobility Model: A model that includes pause times between changes in destination and speed

Random Direction Mobility Model: A model that forces Mobile Nodes (MNs) to travel to the edge of the simulation area before changing direction and speed

Boundless Simulation Area Mobility Model: A model that converts a two-dimensional
   rectangular simulation area into a Torus-shaped simulation area
Gauss-Markov Mobility Model: A model that uses one tuning parameter to vary
   the degree of randomness in the mobility pattern
A probabilistic version of the Random Walk Mobility Model: A model that utilizes
   a set of probabilities to determine the next position of an MN
City Section Mobility Model: A simulation area that represents streets within a city

8.3 Mobility Patterns

In this section, we highlight the challenges arising due to the mobility of nodes,
which the routing algorithms have to deal with. The different mobility patterns and
their importance in ad hoc networks are discussed.

8.3.1 Need for Characterization of Mobility

Efficient operation of the routing protocols is possible only if the correct network
topology information is available to the nodes in the network. So, with this information, the packets could be forwarded correctly between a sender and a receiver.
For the nodes to have a correct topological view of the network, it is required
that the nodes have reasonably correct location information of the nodes in their
neighborhood. If the nodes are stationary, then their location information remains
unchanged, which leads to a correct topological view of the network, and routing
should work well as it does in the wire-line networks. But in ad hoc networks,
the nodes move to different locations over time, thus creating a different network
topology. Almost all well-known routing protocols are shown to perform poorly for
a network where the topology is changing at random.

   With the realization of the importance of node mobility in the routing process
of ad hoc networks, quite a lot of work has been done in mobility characterization of
the mobile nodes. This mobility characterization research is attempting to quantify
the randomness in the mobility of the nodes. Most of the research in this area of
mobility characterization has, however, been toward mobility characterization of
individual nodes. Considering individual node mobility is of minor significance in
ad hoc networks as the nodes in the ad hoc network show cohesive properties. This
observation is directly related to the very existence of ad hoc networks, that is, to
support group collaboration and group activities.

   The effect of movement of neighboring nodes on a movement of an individual node was reported. In this chapter, the importance of treating the mobility
of mobile nodes as a group or the network as a whole, over the mobility model
followed by an individual are described. But the chapter does not throw any light
on how the group mobility information can be used in improving the performance

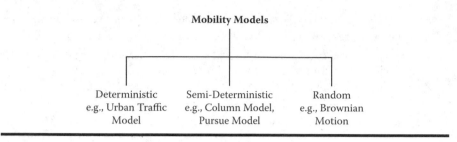

Figure 8.1   Classification of mobility patterns.

of routing algorithms. The major thrust of most of the routing algorithms proposed so far merely has been an attempt to mitigate the effects arising due to mobility that is either by flooding or by letting the nodes in the network constantly update each other on the latest neighborhood information. Either way, it leads to increased overhead. But what most of the algorithms so far have overlooked is the main cause that leads to this instability in the topology, that is, the mobility of the nodes.

8.3.2 Classification of Mobility Patterns

Most of the prior research on ad hoc networks, however, assumed that the nodes follow the Random Mobility Model. And this random model was simulated as "pause and move," with the next move's direction and speed being independent of the previous move made by the nodes. However, it is reasonable to expect a high degree of correlation between future and past movement speeds and direction. Thus, although each node acts separately from all other entities, its decisions for the next move are somehow affected by the context it is in and by the movements surrounding it. And if research could extract this dependence factor, it could bring some amount of predictability into the next movement which the node will make.

The mobility patterns are classified into three patterns as *deterministic* (highly predictable random motion), *semideterministic* (not so predictable random motion), and *random*. (See Figure 8.1.)

8.3.2.1 Deterministic Mobility Model

The Deterministic Mobility Model describes the most predictable type of motion and is the most simplistic of all mobility models. For example, if the mobile nodes were moving in a straight line (i.e., following a Deterministic Model), then the deviation of the direction vectors associated with any two positions would be zero, as the mobile nodes continue to move in the same direction.

A sample scenario resembling a Deterministic Mobility Model would be cars moving in an urban traffic area, where the speed of the cars is restricted and the direction in which the cars can move is also predefined, that is, either in a straight

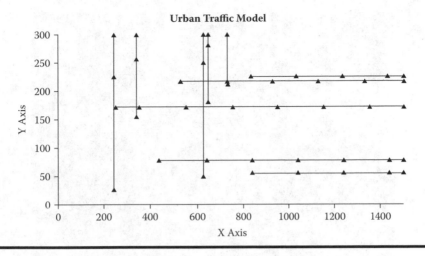

Figure 8.2   A simple scenario for an urban traffic model.

line or turning only at cross lights. This scenario is depicted in Figure 8.2, which is called the "Urban Traffic Model."

8.3.2.2 Semideterministic Mobility Pattern

Now that the deterministic mobility pattern has been described, the scenario in which the mobility pattern followed by the nodes is not so deterministic is discussed briefly. Consider, for example, a battalion of battle tanks marching ahead. Here, the path followed by each tank is not specified, but they do move in a general direction (i.e., toward the war front). Even though the individual tanks do not have a specified direction, we can see a general pattern of a column evolving out of it. Such a mobility pattern is termed a "Column Model." Figure 8.3 shows a pictorial representation of the Column Model, where the ants can be treated as tanks moving in the general direction of the goal in a column.

In this case, the deviation between the direction vectors of two positions can range from −90° to +90° depending on the width of the column as shown by the deviation ϕ in Figure 8.4.

This deviation also signifies the amount of nondeterminism (i.e., the more the average deviation, the more nondeterministic or unpredictable the mobility pattern). That is, with the increase in column width, the maximum possible deviation also increases. The assumption made here is that if nodes are following a Column Model, the nodes will pursue a general direction and not just wander 180° degrees in the opposite direction of the destination. In our view, by varying the deviation (i.e., ϕ in the Semideterministic Model) we can generate various types of mobility patterns, for example the pattern followed by the tanks in the battlefield.

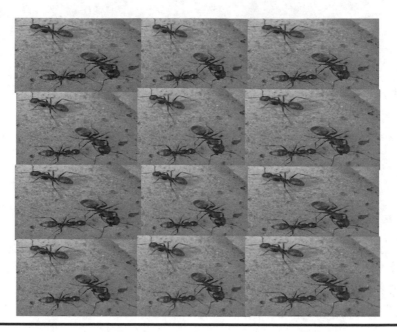

Figure 8.3   The column model.

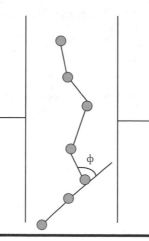

Figure 8.4   Mobile node following a column model.

8.3.2.3 Random Mobility Pattern

Now consider the random motion as in Figure 8.5. This motion is totally stateless, that is, the future movement here is completely independent of the past movement and hence there are no bounds imposed on the max deviation which the nodes can take up for their next movement. And this randomness in choosing the next direction vector renders this type of motion completely unpredictable.

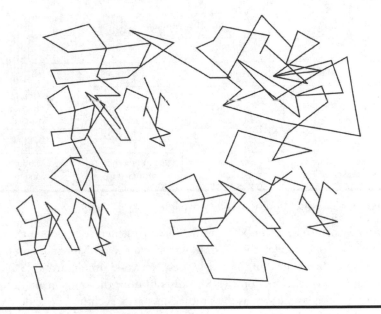

Figure 8.5 Node in a random motion.

8.4 Mobility Models for Mobile Ad Hoc Networks

In this section, we provide categorization for various mobility models into several classes based on their specific mobility characteristics. For some mobility models, the movement of a mobile node is likely to be affected by its movement history. We refer to this type of mobility model as a "mobility model with temporal dependency." In some mobility scenarios, the mobile nodes tend to travel in a correlated manner. We refer to such models as "mobility models with spatial dependency." Another class is the mobility model with geographic restriction, where the movement of nodes is bounded by streets, freeways, or obstacles. The categorization of mobility models is shown in Figure 8.6.

8.4.1 Random-Based Mobility Model

In Random-Based Mobility Models, the mobile nodes move randomly and freely without restrictions. To be more specific, the destination, speed, and direction are all chosen randomly and independently of other nodes. This kind of model has been used in many simulation studies. We have chosen the very commonly used Random Waypoint Model in simulation work.

8.4.1.1 Random Waypoint Model

The Random Waypoint Model was first proposed by Johnson and Maltz. Soon, it became a "benchmark" mobility model to evaluate the MANET routing protocols,

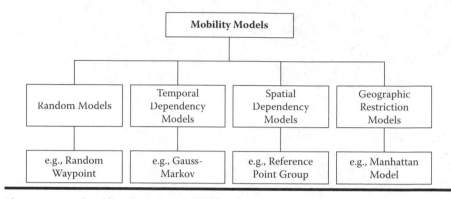

Figure 8.6 Classification of mobility models.

because of its simplicity and wide availability. To generate the node trace of the Random Waypoint Model, the *setdest* tool from the CMU Monarch group may be used. This tool is included in the widely used Network Simulator 2 (NS-2).

In the Network Simulator 2 (NS-2) distribution, the implementation of this mobility model is as follows: as the simulation starts, each mobile node randomly selects one location in the simulation field as the destination. It then travels toward this destination with constant velocity chosen uniformly and randomly from $[0,V_{max}]$, where the parameter V_{max} is the maximum allowable velocity for every mobile node. The velocity and direction of a node are chosen independently of other nodes. Upon reaching the destination, the node stops for a duration defined by the "pause time" parameter T_{pause}. If $T_{pause} = 0$, this leads to continuous mobility. After this duration, it again chooses another random destination in the simulation field and moves toward it. The whole process is repeated again and again until the simulation ends. As an example, the movement trace of a node is shown in Figure 8.7.

In the Random Waypoint Model, V_{max} and T_{pause} are the two key parameters that determine the mobility behavior of nodes. If the V_{max} is small and the T_{pause} is long, the topology of the ad hoc network becomes relatively stable.

On the other hand, if the node moves fast (i.e., the V_{max} is large) and the T_{pause} is small, the topology is expected to be highly dynamic. By varying these two parameters, especially the V_{max} parameter, the Random Waypoint Model can generate various mobility scenarios with different levels of nodal speed. Therefore, it seems necessary to quantify the nodal speed.

Intuitively, one such notion is average node speed. If we could assume that T_{pause}, $= 0$, considering that V_{max} is uniformly and randomly chosen from $[0,V_{max}]$, we can easily find that the average nodal speed is 5.0 V_{max}. However, in general, the pause time parameter should not be ignored. In addition, it is the relative speed of two nodes that determines whether the link between them breaks or forms, rather than their individual speeds. Thus, average node speed seems not to be the appropriate metric to represent the notion of nodal speed.

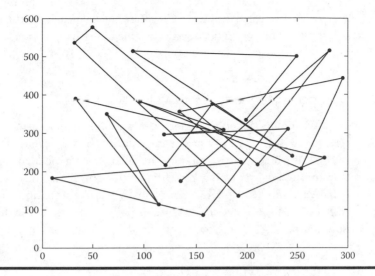

Figure 8.7   Node movement in the Random Waypoint Model.

Johansson et al. took a further step and proposed the mobility metric to capture and quantify this nodal speed notion. The measure of relative speed between node i and j at time t is

$$RS(i,j,t) = \left| V_i(t) - V_j(t) \right| \qquad (8.1)$$

Then, the mobility metric is M calculated as the measure of relative speed averaged over all node pairs and over all time. The formal definition is as follows:

$$\bar{M} = \frac{1}{|i,j|} \sum_{i=1}^{N} \sum_{j=i+1}^{N} \frac{1}{T} \int_0^T RS(i,j,t)dt \qquad (8.2)$$

where $|i,j|$ is the number of the distinct node pair (i,j), n is the total number of nodes in the simulation field (i.e., the ad hoc network), and T is the simulation time.

Using this mobility metric, we are able to roughly measure the level of nodal speed and differentiate the different mobility scenarios based on the level of mobility. Bai, Sadagopan, and Helmy define another mobility metric's average relative speed in a similar way. The experiments show that the average relative speed linearly and monotonically increases with the maximum allowable velocity.

8.4.1.2 Limitations of the Random Waypoint Model

The Random Waypoint Model and its variants are designed to mimic the movement of mobile nodes in a simplified way. Because of their simplicity of implementation and analysis, they are widely accepted. However, they may not adequately capture certain mobility characteristics of some realistic scenarios, including temporal dependency, spatial dependency, and geographic restriction.

8.4.1.2.1 Temporal Dependency of Velocity

In the Random Waypoint Model, the velocity of a mobile node is a memoryless random process, that is, the velocity at the current epoch is independent of the previous epoch. Thus, some extreme mobility behavior, such as sudden stopping, sudden accelerating, and sharp turning, may frequently occur in the trace generated by the Random Waypoint Model. However, in many real-life scenarios, the speed of vehicles and pedestrians will accelerate incrementally. In addition, the direction change is also smooth.

8.4.1.2.2 Spatial Dependency of Velocity

In the Random Waypoint Model, the mobile node is considered an entity that moves independently of other nodes. This kind of mobility model is classified as an entity mobility model. However, in some scenarios, including battlefield communication and museum touring, the movement pattern of a mobile node may be influenced by a certain specific "leader" node in its neighborhood. Hence, the mobility of various nodes is indeed correlated.

8.4.1.2.3 Geographic Restrictions of Movement

In the Random Waypoint Model, the mobile nodes can move freely within the simulation field without any restrictions. However, in many realistic cases, especially the applications used in urban areas, the movement of a mobile node may be bounded by obstacles, buildings, streets, or freeways.

The Random Waypoint Model and its variants fail to represent some mobility characteristics likely to exist in mobile ad hoc networks. Thus, several other mobility models were proposed.

8.4.2 Temporal Dependency Models

The mobility of a node may be constrained and limited by the physical laws of acceleration, velocity, and rate of change of direction. Hence, the current velocity of a mobile node may depend on its previous velocity. Thus, the velocities of single node at different time slots are "correlated."

This mobility characteristic is called the *temporal dependency* of velocity. The memoryless nature of the Random Waypoint Model render inadequate to capture this temporal dependency behavior. As a result, various mobility models considering temporal dependency are proposed. The important mobility model Gauss-Markov is discussed in detail.

8.4.2.1 Gauss-Markov Mobility Model

The Gauss-Markov Mobility Model was first introduced by Liang and Haas [28] and has been widely utilized. In this model, the velocity of the mobile node is assumed to be correlated over time and modeled as a Gauss-Markov stochastic process. In a two-dimensional simulation field, the Gauss-Markov stochastic process can be represented by the following equations:

$$\overline{V_t} = \overline{\alpha} \mathrm{o} \overline{V_{t-1}} + (1 - \overline{\alpha}) \mathrm{o} \overline{\nu} \sqrt{1 - \overline{\alpha}^2} \, \mathrm{o} \overline{W_{t-1}} \tag{8.3}$$

where $V_t = [v_t^x, v_t^y]^T$ and $V_{t-1} = [v_{t-1}^x, v_{t-1}^y]^T$ are the velocity at time t and time $t-1$, respectively. $W_t = [w_{t-1}^x, w_{t-1}^y]^T$ is the uncorrelated random Gaussian process with mean and variance σ^2, and $\overline{\alpha} = [\alpha^x, \alpha^y]^T$, $\overline{\nu} = [\nu^x, \nu^y]$, and $\overline{\sigma} = [\sigma^x, \sigma^y]^T$ are the vectors that represent the memory level, asymptotic mean, and asymptotic standard deviation, respectively.

For the sake of simplicity, we may represent the general form of equation (8.3) in a two-dimensional field as follows:

$$v_t^y = \alpha v_{t-1}^y + \left(1 - \alpha\right) v^y + \sigma^x \sqrt{1 - \alpha^2} \, w_{t-1}^y \tag{8.4}$$

$$v_t^x = \alpha v_{t-1}^x + \left(1 - \alpha\right) v^x + \sigma^x \sqrt{1 - \alpha^2} \, w_{t-1}^x \tag{8.5}$$

When the node is going to travel beyond the boundaries of the simulation field, the direction of movement is forced to flip 180 degrees. This way, the nodes remain within the boundary of the simulation field.

Based on these equations, we observe that the velocity $V_t = [v_t^x, v_t^y]^T$ of the mobile node at time slot t is dependent on the velocity $V_{t-1} = [v_{t-1}^x, v_{t-1}^y]^T$ at time slot $t-1$. Therefore, the Gauss-Markov Model is a temporally dependent mobility model whereas the degree of dependency is determined by the memory level parameter α. α is a parameter to reflect the randomness of the Gauss-Markov process. By tuning this parameter, the model is capable of duplicating different kinds of mobility behaviors in various scenarios.

1. If the Gauss-Markov Model is memoryless (i.e., $\alpha = 0$), equations (8.4) and (8.5) are

$$v_t^x = \mathsf{v}^x + \sigma^x w_{t-1}^x \tag{8.6}$$

$$v_t^y = \mathsf{v}^y + \sigma^y w_{t-1}^y \tag{8.7}$$

where the velocity of the mobile node at time slot t is only determined by the fixed drift velocity $\bar{v} = [v^x, v^y]^T$ and the Gaussian random variable $W_t = [w_{t-1}^x, w_{t-1}^y]^T$. Obviously, the model described in equations (8.6) and (8.7) is the Random Walk Model.

2. If the Gauss-Markov Model has strong memory (i.e., $\alpha = 1$), equations (8.4) and (8.5) are

$$v_t^x = v_{t-1}^x \tag{8.8}$$

$$v_t^y = v_{t-1}^y \tag{8.9}$$

where the velocity of the mobile node at time slot t is exactly the same as its previous velocity. In the nomenclature of vehicular traffic theory, this model is called the "Fluid Flow Model."

3. If the Gauss-Markov Model has some memory (i.e., $0 < \alpha < 1$), the velocity at the current time slot is dependent on both its velocity $V_{t-1} = [v_{t-1}^x, v_{t-1}^y]^T$ at time t–1 and a new Gaussian random variable $W_t = [w_{t-1}^x, w_{t-1}^y]^T$. The degree of randomness is adjusted by the memory level parameter α. As α increases, the current velocity is more mainly affected by its previous velocity. Otherwise, it will be mainly affected by the Gaussian random variable. In the Gauss-Markov Model, the temporal dependency plays a key role in determining the mobility behavior.

8.4.3 *Spatial Dependency Models*

In the Random Waypoint Model, a mobile node moves independently of other nodes, that is, the location, speed, and movement direction of a mobile node are not affected by other nodes in the neighborhood. As previously mentioned, these models do not capture many realistic scenarios of mobility. For example, in the case of a vehicle on a freeway attempting to avoid a collision, the vehicle's speed cannot exceed that of the vehicle ahead of it. Moreover, in some targeted MANET applications including disaster relief and battlefields, team collaboration among users exists and the users are likely to follow the team leader. Therefore, the mobility of a mobile node could be influenced by other neighboring nodes. Because the velocities of different nodes are "correlated" in space, we call this characteristic the Spatial Dependency of Velocity. The Reference Point Group Mobility Model (RPGM) will be discussed in detail.

8.4.3.1 Reference Point Group Mobility (RPGM) Model

In line with the observation that the mobile nodes in MANET tend to coordinate their movement, the Reference Point Group Mobility (RPGM) Model is proposed. One example of such mobility is that a number of soldiers may move together in a group or platoon. Another example is during disaster relief, where various rescue crews (e.g., firefighters, police officers, and medical assistants) form different groups and work cooperatively.

In the RPGM Model, each group has a center, which is either a logical center or a group leader node. For the sake of simplicity, we assume that the center is the group leader. Thus, each group is composed of one leader and a number of members. The movement of the group leader determines the mobility behavior of the entire group. The respective functions of group leaders and group members are described as follows.

8.4.3.1.1 Group Leader

The movement of the group leader at time t can be represented by motion vector V^t_{group}. Not only does it define the motion of the group leader itself, but also it provides the general motion trend of the whole group. Each member of this group deviates from this general motion vector V^t_{group} by some degree. The motion vector V^t_{group} can be randomly chosen or carefully designed based on certain predefined paths.

8.4.3.1.2 Group Members

The movement of group members is significantly affected by the movement of the group leader. For each node, mobility is assigned with a reference point that follows the group movement. Upon this predefined reference point, each mobile node

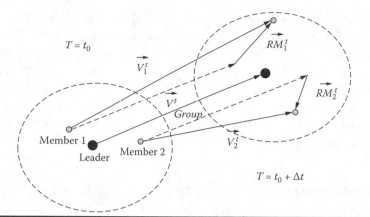

Figure 8.8 **Node movement in the RPGM Model with two snapshots at a time.**

could be randomly placed in the neighborhood. Formally, the motion vector of group member i at time t, V_{group}^t , can be described as

$$V_i^t = V_{group}^t + R\,M_i^t \tag{8.10}$$

where the motion vector $R\,M_i^t$ is a random vector deviated by group member i from its own reference point. The vector $R\,M_i^t$ is an independent identically distributed (IID) random process whose length is uniformly distributed in the $[0, r_{max}]$ (where r_{max} is maximum allowed distance deviation) and whose direction is uniformly distributed in the interval $[0, 2\pi]$. Figure 8.8 illustrates an example of the RPGM Model. In Figure 8.8, V_{group}^t is the motion vector for the group leader; it is also the motion vector for the whole group. $R\,M_i^t$ is the random deviation vector for the whole group. $R\,M_i^t$ is the random deviation vector for group member i, and the final motion vector of group member i is represented by vector V_i^t .

With the appropriate selection of predefined paths for the group leader and other parameters, the RPGM Model is able to emulate a variety of mobility behaviors. For example, the RPGM Model is able to represent various mobility scenarios, including the following:

1. *In-Place Mobility Model*: The entire field is divided into several adjacent regions. Each region is exclusively occupied by a single group. One such example is battlefield communication.

2. *Overlap Mobility Model*: Different groups with different tasks travel on the same field in an overlapping manner. Disaster relief is a good example.
3. *Convention Mobility Model*: This scenario emulates the mobility behavior at a conference. The area is also divided into several regions, and some groups are allowed to travel between regions.

In the mobility vector framework, an extension of the RPGM Model is proposed. In this framework, authors point out that many realistic mobility scenarios could be modeled and generated with this framework by properly choosing the checkpoints along the preferred motion path of the group leader.

If those checkpoints can reflect the motion behavior in realistic scenarios, then the Mobility Vector Model provides a general and flexible framework for describing and modeling mobility patterns. However, in practice, it is not a trivial task to generate those checkpoints. In the RPGM Model, the vector $R\,M_i^t$ indirectly determines how much the motion of group members deviates from their leader. So, we are not able to generate the various mobility scenarios with different levels of spatial dependency by simple adjustment of model parameters. To solve this problem, a modified version of the RPGM Model is proposed. The movement can be characterized as follows:

$$\left|V_{member}\left(t\right)\right| = \left|V_{leader}\left(t\right)\right| + random() * SDR * \max\_speed \qquad (8.11)$$

$$\theta_{member}\left(t\right) = \theta_{leader}\left(t\right) + random() * ADR * \max\_angle \qquad (8.12)$$

where $0 < SDR$, $1 < ADR$, SDR is the Speed Deviation Ratio, and ADR is the Angle Deviation Ratio. SDR and ADR are used to control the deviation of the velocity (magnitude and direction) of group members from that of the leader. By simply adjusting these two parameters, different mobility scenarios can be generated.

Because of the inherent characteristics of spatial dependency between nodes, the RPGM Model is expected to behave differently from the Random Waypoint Model. The RPGM Model incurs less link breakage and achieves a better performance for various routing protocols than the Random Waypoint Model.

8.4.4 Geographic Restriction Model

Another limitation of the Random Waypoint Model is the unconstraint motion of the mobile node. Mobile nodes, in the Random Waypoint Model, are allowed to move freely and randomly anywhere in the simulation field. However, in most real-life applications, we observe that a node's movement is subject to its environment.

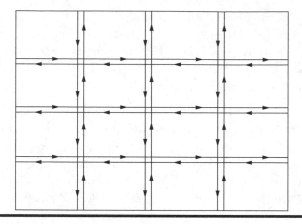

Figure 8.9   Manhattan Mobility Model.

In particular, the motions of vehicles are bounded to the freeways or local streets in the urban area, and on campus the pedestrians may be blocked by the buildings and other obstacles. Therefore, the nodes may move in a pseudorandom way on predefined pathways in the simulation field. Some recent works address this characteristic and integrate the paths and obstacles into mobility models. This kind of mobility model is called a "mobility model with geographic restriction." The Manhattan Mobility Model is a popular geographic restriction model which we have used in our simulation. The following section describes the Manhattan Mobility Model in detail.

One simple way to integrate geographic constraints into the mobility model is to restrict the node movement to the pathways in the map. The map is predefined in the simulation field and utilizes a random graph to model the map of the city. This graph can be either randomly generated or carefully defined based on a certain map of a real city. The vertices of the graph represent the buildings of the city, and the edges model the streets and freeways between those buildings.

Initially, the nodes are placed randomly on the edges of the graph. Then for each node a destination is randomly chosen and the node moves toward this destination through the shortest path along the edges. Upon arrival, the node pauses for T_{pause} time and again chooses a new destination for the next movement. This procedure is repeated until the end of the simulation. Unlike the Random Waypoint Model where the nodes can move freely, the mobile nodes in this model are only allowed to travel on the pathways. However, because the destination of each motion phase is randomly chosen, a certain level of randomness still exists for this model. So, in this graph-based mobility model, the nodes are traveling in a pseudorandom fashion on the pathways.

Similarly, in the Freeway Mobility Model and Manhattan Mobility Model, the movement of the mobile node is also restricted to the pathway in the simulation field. Figure 8.9 illustrates the maps used for the Manhattan Models.

8.5 Summary

Much of the initial research was based on using the Random Waypoint Model as the underlying mobility model and CBR traffic consisting of randomly chosen source destination pairs as the traffic pattern. Routing protocols like DSR, DSDV, AODV, and temporally ordered routing algorithm (TORA) were mainly evaluated based on the packet delivery ratio (ratio of the number of packets received to the number of packets sent) and routing overhead (number of routing control packets sent) metrics, and end-to-end delay was not considered. On-demand protocols such as DSR and AODV performed better than table-driven ones such as DSDV at high mobility rates, whereas DSDV performed quite well at low mobility rates. It was observed that DSR outperforms AODV in less demanding situations, whereas AODV outperforms DSR in cases of heavy traffic load and high mobility. However, the routing overhead of DSR was found to be less than that of AODV. In the above studies, focus was given on performance evaluation, whereas parameters investigated in the mobility model were change of maximum velocity and pause time. Not only the Random Waypoint Mobility Model was used, but also other mobility models such as RPGM, Gauss-Markov, and Manhattan in evaluation of the performance of routing protocols.

8.6 Problems

8.1 Why are mobility models needed for a wireless environment?
8.2 Give the classification of mobility models.
8.3 Justify the need for characterizations of mobility.
8.4 Discuss the classification of mobility patterns.
8.5 Explain the Column Model with a suitable example.
8.6 Describe the Random Waypoint Mobility Model with an example.
8.7 What are the limitations of the Random Waypoint Mobility Model?
8.8 What are temporal dependency models? Explain.
8.9 Describe the Gauss-Markov Mobility Model with a neat diagram.
8.10 What are spatial dependency models? Explain with an example.
8.11 Explain the Geographic Restriction Model with an illustration.

Bibliography

F. Bai, N. Sadagopan, and A. Helmy, Important: A framework to systematically analyze the impact of mobility on performance of routing protocols for ad hoc networks, in Proceedings of the IEEE Information Communications Conference (INFOCOM 2003), San Francisco, April 2003.

T. G. Basavaraju, C. Puttamadappa, and S. K. Sarkar, Influence of mobility models on the performance of routing protocols in mobile ad hoc networks, *Journal of Computer Science*, 7(6), June 2007.

T. G. Basavaraju, K. C. Chetan, C. Puttamadappa, and S. K. Sarkar, Affect of mobility models on the routing protocol performance in mobile ad hoc networks, in Proceedings of the Third International Conference ObCom-2006, Mobile, Ubiquitous & Pervasive Computing, Vellore Institute of Technology, Vellore, Tamil Nadu, India, Vol. 2, pp. 101–107, Dec. 2006.

L. Breslau, D. Estrin, K. Fall, S. Floyd, J. Heidemann, A. Helmy, P. Huang, S. McCanne, K. Varadhan, Y. Xu, and H. Yu, Advances in network simulation, in *IEEE Computer*, 33(5): 59–67, May 2000.

J. Broch, D. A. Maltz, D. B. Johnson, Y-C. Hu, and J. Jetcheva, A performance comparison of multi-hop wireless ad hoc network routing protocols, in Proceedings of the Fourth Annual ACM/IEEE International Conference on Mobile Computing and Networking, ACM, Oct. 1998.

T. Camp, J. Boleng, and V. Davies, A survey of mobility models for ad hoc network research, *Wireless Communication and Mobile Computing* (WCMC), special issue on mobile ad hoc networking: Research, trends and applications, 2(5):483–502, 2002.

S. R. Das, C. E. Perkins, and E. M. Royer, Performance comparison of two on-demand routing protocols for ad hoc networks, in INFOCOM, March 2000.

S. R. Das, R. Castaneda, J. Yan, and R. Sengupta, Comparative performance evaluation of routing protocols for mobile, ad hoc networks, in 7th International Conference on Computer Communications and Networks (IC3N), Oct. 1998, pp. 153–161.

X. Hong, T. Kwon, M. Gerla, D. Gu, and G. Pei, A mobility framework for ad hoc wireless networks, in the ACM Second International Conference on Mobile Data Management (MDM), Jan. 2001.

X. Hong, M. Gerla, G. Pei, and C C. Chiang, A group mobility model for ad hoc wireless networks, in Proceedings of ACM/IEEE MSWiM'99, Seattle, WA, Aug. 1999, pp. 53–60.

Y-C. Hu and D. B. Johnson, Caching strategies in on-demand routing protocols for wireless ad hoc networks, in Proceedings of the Sixth Annual International Conference on Mobile Computing and Networking (MobiCom 2000), ACM, Boston, MA, Aug. 2000.

P. Johansson, T. Larsson, N. Hedman, B. Mielczarek, and M. Degermark, Scenario-based performance analysis of routing protocols for mobile ad-hoc networks, in Proceedings of the International Conference on Mobile Computing and Networking (MobiCom'99), 1999, pp. 195–206.

D. B. Johnson, D. A. Maltz, and J. Broch, DSR: The dynamic source routing protocol for multi-hop wireless ad hoc networks, in *Ad hoc networking*, C. Perkins, Ed., Addison-Wesley, Reading, MA, 2001, pp. 139–172.

D. Lam, D. C. Cox, and J. Widom, Teletraffic modeling for personal communication services, *IEEE Communications*, 35(2):79–87, Oct. 1999.

B. Liang and Z. J. Haas, Predictive distance-based mobility management for PCS networks, in Proceedings of the IEEE Information Communications Conference (INFOCOM 1999), April 1999.

T. Liu, P. Bahl, and I. Chlamtac, Mobility modeling, location tracking, and trajectory prediction in wireless ATM networks, *IEEE J. Sel. Areas in Commun.*, 16(6):922–936, Aug. 1998.

J. G. Markdoulidakis, G. L. Lyberopoulos, D. F. Tsirkas, and E. D. Sykas, Mobility modeling in third-generation mobile telecommunication systems, *IEEE Personal Communications*, 41–56, Aug. 1997.

A. B. McDonald and T. Znati, Predicting node proximity in ad-hoc networks: A least overhead adaptive model for selecting stable routes, in Proceedings of the ACM/MobiHoc2000, Boston, MA, Aug. 2000, pp. 29–33.

A. B. McDonald and T. Znati, A path availability model for wireless ad-hoc networks, in Proceedings of the IEEE Wireless Communications and Networking Conference 1999 (WCNC'99), New Orleans, LA, Sept. 21–24, 1999.

The Network Simulator 2 (NS-2), http://www.isi.edu/nsnam/ns.

V. D. Park and M. S. Corson, Temporally-ordered routing algorithm (TORA) version 1: Functional specification, Internet draft, http://tools.ietf.org/html/draft-ietf-manet-tora-spec-01, Aug. 1998.

G. Pei, M. Gerla, X. Hong, and C-C.-Chiang, A wireless hierarchical protocol with group mobility, in Proceedings of IEEE WCNC'99, New Orleans, LA, Sept. 1999.

C. Perkins, Ad hoc on demand distance vector (AODV) routing, Internet draft, http://moment.cs.ucsb.edu/pub/draft-perkins-manet-aodvbis-00.txt.

C. E. Perkins and P. Bhagwat, Highly dynamic destination sequenced distance vector routing (DSDV) for mobile computers, in ACM SIGCOMM, 234–244, 1994.

D. S. Tan, S. Zhou, J. Ho, J. S. Mehta, and H. Tanabe, Design and evaluation of an individually simulated mobility model in wireless ad hoc networks, in Proceedings of the Communication Networks and Distributed Systems Modeling and Simulation Conference 2002, San Antonio, TX.

J. Tian, J. Hahner, C. Becker, I. Stepanov, and K. Rothermel, Graph-based mobility model for mobile ad hoc network simulation, in the Proceedings of the 35th Annual Simulation Symposium, in cooperation with the IEEE Computer Society and ACM, San Diego, CA, April 2002.

Chapter 9

Cross-Layer Design Issues for Ad Hoc Wireless Networks

9.1 Introduction

Traditionally, network protocols are divided into several independent layers. Each layer is designed separately, and the interaction between layers is performed through a well-defined interface. The main advantage of this type of approach is architectural flexibility. One implementation of a layer can be seamlessly replaced with another implementation. For example, if an old protocol has to be replaced with a newer one, there is no need to modify the rest of the network stack. Apart from the flexibility, in wired networking applications this approach to physical, Medium Access Control (MAC), and network layers does not impose a performance penalty: separate wired links are independent of each other and do not affect each other's performance. It is different for wireless networks. One of the differences is that links can no longer be viewed as separate entities whose performances are independent of each other. Each wireless transmission can be heard by nodes in the neighborhood and is perceived by them as interference. As a consequence of the interdependency, there is a need for a more complex medium access mechanism. On the one hand, this mechanism should be able to control the amount of interference experienced by receivers. On the other hand, it should exploit spatial reuse and, in certain cases, enforce concurrent transmissions, to maximize the performance. Because medium access protocol controls the amount of interference in the network, it influences the performance of the physical layer. If the total amount of interference at a receiver during reception of a packet is high, the physical layer should decrease the transmission rate to cope with it. On the contrary, if the interference is low, the physical layer should benefit from the conditions and transmit with a high rate. This is again contrary to the approach in a wired network where any two concurrent transmissions always cause a collision, and there is no concept of rate adaptation as a function of existing traffic. Another degree of freedom that does not exist in the physical layers of wired networks is power control. The higher the packet transmission power is, the higher the received power is, hence the higher the transmission rate. However, if we increase the transmission power, at the same time we increase the interference to other nodes. Power control is thus tightly coupled with both the physical and medium access layers. Perhaps the least obvious interdependency is between the physical layer, medium access, and routing. The wireless medium allows connections between any two nodes in a region: the further away the nodes are, the lower the achievable communication rate will be. If the direct connection can be established only with a very low rate, the routing protocol may decide to relay over intermediate nodes. This way, a single packet will be transmitted on several short links: it will consume the resources of several nodes, but for a shorter period of time, because each intermediate transmission will be on a higher rate. It is easy to see that a change in routing policy may change the performance of the layers underneath; namely, if a routing protocol decides to relay one flow over several hops, instead of transmitting directly from a source to a destination, we will have a number of short links on the route. Some of these links will be active at the same time, and this will create more interference on other nodes than a single long hop. At the same time, these shorter links will be more resistant to interference than the single long link. From the above examples, it is clear that a change in a protocol of one layer will affect the performance of other layers. The independent design of different layers may yield a grossly suboptimal network. This chapter analyzes a cross-layer network design that will yield jointly optimal physical, medium access, and routing layers for wireless ad hoc networks.

9.2 Cross-Layer Design Principle

One of the major components in the success of the Internet is the layered Open System Interconnection (OSI) architecture. The modularity achieved through layering leads to better understanding of the abstract functionality of layers and thus enables better understanding of the overall system. This led to rapid growth in the development of a number of applications that drive the Internet. The fact that the same lower layers may be reused for every application decreases the development cost of the application and enhances the utility of the network architecture.

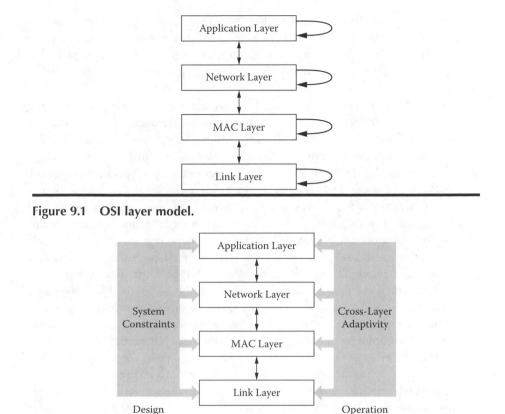

Figure 9.1   OSI layer model.

Figure 9.2   Cross-layer model.

Layering simplifies network design and leads to robust scalable protocols in the Internet. Layering, however, suffers from suboptimality and inflexibility. Layering is suboptimal because each layer has insufficient information about the network because it does not allow the sharing of information among the layers.

The interface between the layers is static and independent of individual network constraints and applications. Layering is inflexible because the developer of a new application has to rely solely on the functionality of the lower layers. According to the discussion presented in the previous section, the layers in a wireless network must coordinate and adapt with the change in the state of the wireless network. This is the motivation behind the cross-layer paradigm for protocol design in wireless networks. (See Figure 9.1 and Figure 9.2.)

The cross-layer design of a protocol stack enables layers to exchange state information to adapt and optimize the performance of the network. The sharing of information enables each layer to have a global picture of the constraints and characteristics of the network, leads to better coordination, and enables them to make decisions that would jointly optimize the performance of the network. The cross-layer principle further requires that the protocols must be developed not in isolation but in an integrated and hierarchical framework so as to take advantage of the interdependencies between the protocols. These interdependencies are related to the adaptively at each layer, the system constraints, and the requirements of the application.

In cross-layer architecture, a MAC layer may adapt its scheduling based on the link quality and interference such that the performance constraints of the application are satisfied. Thus, a MAC layer needs to have information about the link characteristics from the link (lower) layer and the performance constraints from the application (upper) layer. Similarly, an adaptive cross-layer routing protocol may choose the routes based on the information about the link characteristics and MAC scheduling policy to meet performance requirements. It is important to understand that to adapt to a change in the network, a layer must first try local adaptation and inform the upper layer about the change only if the local adaptation does not work. This is because the timescale of changes at lower levels is much lower than the timescale of changes at the upper layers. For example, the SINR of a link may change much more rapidly than the position of a node. So when the quality of a link degrades, the link layer must first try to adapt to the change, possibly by increasing the transmit power or using better coding. This would temporarily solve the problem if the change in SINR is due to a random fluctuation and the SINR of the link would later be restored. However, if the SINR of the link does not improve for a long time, then the link layer realizes that this degradation may be due to a change of the topology (the other node might have moved away), so it informs the network layer that something has gone wrong with the link. The network layer then recalculates the routes using this information. According to OSI architecture, the main issues that need to be considered while designing a cross-layer protocol stack are as follows:

- What information should be exchanged across protocol layers, and how should that information be adapted?
- How should the global system constraints and characteristics be factored into the protocol designs at each layer?

9.3 Proposals Involving Cross-Layer Design

In the previous section, we emphasized the importance of cross-layer design and presented broad principles behind the cross-layer design. In this section, we present joint physical-MAC layer design for networks under Rayleigh fading using cross-layer design to improve the performance of wireless networks. The wireless channel is characterized by random time-varying fading, which dictates the SINR of the received signal. If the channel is in the state of deep fade, then the SINR of a received packet is very low and the receiver is not able to decode the message correctly. Thus, the packet has to be dropped by the receiver although no collision had occurred. In a layered protocol stack, the upper layers are unaware of the state of the wireless channel. As a result, a node may keep on transmitting packets while the channel is in a bad state. This would not only waste the resources of the transmitter but also cause needless interference at the neighboring nodes. A cross-layer design is proposed where the MAC layer predicts the state of the channel to make sure that the nodes do not transmit when the state of the channel is bad. The prediction is based upon an analytical model of a Rayleigh fading channel and utilizes the past signal strength measurements. If the MAC layer predicts that the state of the channel is good, then it defers transmission. The prediction model for the Rayleigh fading channel is incorporated with a Markovian model for Institute of Electrical and Electronic Engineers (IEEE) 802.11 standards MAC to analyze the performance of the proposed approach. The predictability of the Rayleigh fading channel is used to improve performance of the network in the following manner. The node at the receiving end observes the power levels of each received transmission from the receiver. Based on these measurements, the receiver predicts whether the channel would be in good or bad state during the next transmission. If the receiver predicts that the state of the channel is going to be bad, then it informs the sender about the fade and stops transmission of any reply packets to the sender. The receiver may inform the sender about the imminent fade by setting a flag in the acknowledgment (ACK) or clear-to-send (CTS) packet that it transmits to the sender. When the sender receives this notification, then it immediately halts the transmission, calculates the expected fade duration, and schedules future transmissions accordingly. The Network Allocation Vector (NAV) at the neighbors is also updated when they overhear a CTS or ACK whose flag bit is marked. The simulation results obtained indicate the cross-layer implementation performs better than the layer implementation in terms of received signal strength, throughput, and fraction of packets dropped.

9.4 Cross-Layer Design: Is It Worth Applying It?

In the previous sections, we discussed the advantages offered by the cross-layer design and presented some proposals that use cross-layer design. Cross-layer feedback helps achieve better overall system performance in an ad hoc wireless network. Cross-layer feedback can be both ways. Feedback does not need to be between adjacent layers only. Although each layer may help other layers to contribute to the system's performance, redundancy of cross-layer information may be avoided. Cross-layer feedback should be considered while designing ad hoc networks. Fast handoff due to node mobility, adaptive Session Initiation Protocol (SIP)–based application, and domain autoconfiguration were cited as contexts where cross-layer feedback can be useful.

9.5 Cross-Layer Design in Wireless Networks

Access to the radio medium is traditionally considered a problem of the MAC layer. However, it has recently become evident that a traditional layering approach that separates routing, scheduling, flow control, power, and rate control might not be efficient for ad hoc wireless networks. This is primarily due to the interaction of links through interference, which implies that a change in power allocation or schedules on one link can induce changes in the capacities of all links in the surrounding area and changes in the performance of flows that do not use the modified link.

9.5.1 Fundamental Advantages Offered by a Layered Architecture

The layered architecture is not unique to the network protocol stack but is common to many popular systems that have withstood the test of time. The main advantages achieved by layering are modularity and reusability. Modularity implies that each layer may be developed or upgraded independently of other layers. Reusability allows sharing of lower layers by several instances of upper layers, thus reducing the cost of development and deployment of the system. This leads to a rapid proliferation of the layered systems. Although the layered architecture may not be optimal in the theoretical sense, the performance enhancements that it guarantees are longevity of the system and low implementation costs.

Examples of such layered architecture are as follows:

The von Neumann architecture: The von Neumann architecture is at the heart of most computer systems. It classifies the computer system into independent functional units (layers): the memory unit, control unit, arithmetic and logical unit, and input–output unit. The architecture makes it possible that hardware and software for the computer systems may be developed independently. This is one of the major reasons behind the rapid proliferation of computer systems.

Source–channel separation and digital system architecture: Shannon proved that the layers of source compression (source coding) and coding for reliable transmission over a wireless channel (channel coding) may be implemented separately and independently. This implied that each new source of information (and the associated source encoder–decoder) may simply reuse the existing channel encoders–decoders, thus simplifying the implementation. In the same way, new (more efficient) channel encoders–decoders may be designed without worrying about the sources that would be using the channel. This architecture has fueled the rapid development and proliferation of digital communication systems.

The OSI architecture for networking: The OSI architecture and its impact on the development and proliferation of computer networks have been sufficiently discussed in this study.

9.6 Performance Objectives

A cross-layer design of a wireless ad hoc network is essentially an optimization problem with several degrees of freedom. Before proceeding to solving the optimization problem, we have to define the performance objective. Roughly, performance objectives, in the case of wireless ad hoc networks, can be divided into two categories. The first category comprises power-based performance objectives that maximize network lifetime. A typical example is a sensor network. Traffic requirement is low, and the main goal is to maintain a network in operation as long as possible. The second category comprises rate-based performance objectives. For these, the goal is to maximize flow rates. Typical examples are wireless local area networks (LANs), networks of computer peripherals (e.g., wireless replacements for USB or FireWire), or home appliances. The overall power consumption of nodes of such a network is typically much larger than the energy spent on wireless transmissions, hence there is no need to optimize network lifetime, whereas applications, such as video, audio, or data transmissions, may require high data rates.

This chapter has considered the wireless networks with best-effort traffic, focused on rate-based performance metrics, and analyzed the well-known rate-based performance metrics, originally defined in the wired networking context: rate maximization, proportional fairness and max–min fairness, and some of their modifications. Performance objectives for wireless network design can roughly be divided into two categories.

1. Energy-based performance objectives, whose goal is to minimize energy dissipation of a network and increase network lifetime. This category is out of the scope of this chapter.
2. The other category comprises rate-based performance objectives, whose goal is to maximize capacity.

There are three common rate-based performance metrics used in networking. The three most frequently used design objectives for wireless networks are *maximizing total capacity*, *max–min fairness*, and *utility fairness*.

9.6.1 Maximizing Total Capacity

Maximizing total capacity is a performance objective traditionally used in the design of cellular systems. This objective comes from a voice setting where all users have the same capacity need, and the goal is to maximize the total capacity (i.e., the maximum number of calls that can be supported by a network). It is, however, much less suitable for data traffic. It has been known in the world of wired networking that maximizing total capacity is an unfair design objective. The same problem persists in wireless networks. A strategy that maximizes the total capacity is such that a node with the best channel conditions in a given slot should send data. Nodes that are farther away will less frequently satisfy this constraint, but will still have a positive throughput due to the random part of fading. However, if a node is very far away from the base station, its average rate will be very small and essentially it will not be able to communicate. A remedy is found by assigning weights to node rates such that a level of fairness is assured. The implicit assumption in this type of network is that an area with mobile nodes is well covered with base stations, so there is not a great variation in distances from the mobile nodes to the closest base stations. A similar direction was taken for High Date Rate (HDR)–code division multiple access (CDMA) networks.

In the case of multihop wireless networks, variations in the distances between sources and destinations are typically much higher because a node does not talk to the closest base station but to an arbitrary destination in a network. This makes it difficult to remedy fairness with weights, and longer flows risk low or zero throughputs. For example, it has been observed in this context that the unfairness of maximizing the sum of rates persists in wireless networks, and some long-distance flows obtain zero throughputs. A famous rate-based metric, based on the weighted sum of rates and used in multihop wireless settings, is transport capacity.

9.6.2 Max–Min Fairness

In the context of wired networks, max–min fair rate allocation is defined to be the flow rate allocation in which every flow has a bottleneck link. This rate allocation can be obtained using the well-known water-filling algorithm. However, it is not always obvious how to generalize the notion of a bottleneck link and the water-filling approach to an arbitrary problem. A simple example is a wireless ad-hoc network. Even if every flow has a bottleneck link, the allocation may not be max–min fair. In particular, by decreasing the transmit power of some links, one can decrease the interference on the receiver of one of the bottleneck links and thus increase the link's rate. This way, a flow will loosen its bottleneck, and one might be able to increase its rate. It is difficult to define the concepts of a bottleneck link and of the water filling in the given example, and furthermore, it is not obvious if the max–min fair rate allocation can be defined at all in this.

9.6.3 Utility Fairness

A different approach to fairness in wired networking is utility fairness. Every user is assigned a utility which is a function of its rate. The goal of a network is to maximize the sum of utilities of all users. By selecting utility functions, a designer can achieve different trade-offs between efficiency and fairness. It can also be shown that Transmission Control Protocol (TCP) implements a form of utility fairness. Variants of utility fairness are used in existing wireless multihop network protocols.

9.7 Pitfalls of the Cross-Layer Design Approach

9.7.1 Cost of Development

The aim of cross-layer design is to adapt according to the network state so as to optimize the performance of the applications. Some actions of the protocols are highly dependent on the application that the network is designed to support. If the demands of two applications or environments of two instances of wireless network are radically different, then each of them would require a separate set of protocols. Thus, the cross layers would have to be handcrafted for each application and network scenario. For example, consider two networks, N1 and N2. Suppose that in N1 the nodes are battery operated, whereas in N2 the nodes have an infinite source of power (e.g., they are connected to an electrical socket or operated on solar power). So in N1 one of the objectives of the optimization would be to consume minimum energy, whereas in N2 this would not be valid. Although the protocol set of N1 would also work fine in N2, this again would lead to suboptimal performance in N2, to counter the reason why the cross-layer design was developed in the first place. So the handcrafting of protocols for each application and network would lead to high deployment costs and delays.

9.7.2 Performance versus Longevity

The most popular argument in favor of sharing information among layers is that it leads to optimal performance. However, such a gain in performance is of a short-sighted nature. This is because the technologies at each layer (e.g., capabilities of MAC cards or coding schemes at link layers) change rapidly. So with every change of technology, the nature of information that is shared and the actions that are taken would need to be changed. This is against the principle of longevity, which is considered an essential feature of any design. If we include the weights of longevity and cost while evaluating the overall performance of architecture, then the layered architecture may as well outperform the architectures that make aggressive use of the cross-layer approach.

9.7.3 Interaction and Unintended Consequences

The layered architecture allows limited and controlled interactions between the layers such that the job of designing or modifying a protocol at any level is simplified. A cross-layered approach leads to many dependencies between various layers. The designer of a new protocol in a cross-layer system has to understand and take into account the interaction of various layers. In spite of a good understanding, a new protocol may lead to unintended consequences due to the presence of multiple adaptation loops. Such interactions need to study using dependency graphs. Thus, design of a protocol in a cross-layer stack is much more challenging than the task of designing such a protocol in a layered stack.

9.7.4 Stability

As was already mentioned, a cross-layer design leads to several adaptation loops. The complex interaction of these loops may endanger the stability of the system. Although cross-layer design offers tremendous opportunities, at the same time it has several critical disadvantages. These disadvantages may hinder the proliferation of wireless networks. The cross-layer approach should thus be used with caution. There is much functionality, like transmit power control and channel state estimation that are typical of wireless networks and require a cross-layer approach. In such cases where the cross-layer design is necessary, it should be made sure that the implementation is not very aggressive (i.e., does not rely too much on information exchanged among several layers). The placement of functions and the nature of the information exchanged between the layers must be kept at a minimum and must be critically analyzed.

9.8 Summary

Although the cross-layer design may provide performance optimization, it leads to an increased cost of deployment, endangers stability, and would hinder the proliferation and development of wireless networks. So the aggressive use of cross-layer design is not a good idea. A better approach would be to carefully assess the placement of various functionalities within the protocol stack and minimize its dependency on the information from other layers. The approach is illustrated by the example of the design of transmit power control protocols for wireless ad hoc networks.

It should, however, be noted that cross-layer design may be a suitable approach for stand-alone wireless networks that are dedicated to a single application, especially if the task is highly critical, for example surveillance of a nuclear plant, gathering of seismic information for predicting earthquakes, data gathering behind enemy lines, and so forth. In such networks, reliability and performance are more important than cost. Also, because a single application has to be supported, we don't have to worry about issues like interoperability. Thus, aggressive use of cross-layer design could be highly valuable for such task-oriented networks.

9.9 Problems

9.1 Explain the cross-layer design principle.

9.2 Describe the proposals involving cross-layer design for ad hoc networks.

9.3 Discuss the fundamental advantages offered by a layered architecture.

9.4 Explain some of the standard layered architecture with an example.

9.5 How is a performance objective met while designing the layered architecture?

9.6 Discuss the pitfalls of the cross-layer design approach.

Bibliography

D. De Couto et al., A high-throughput path metric for multi-hop wireless routing, Proc. Mobicom 2003.

Y. Fang and A. B. McDonald, Cross-layer performance effects of path coupling in wireless ad hoc networks: power and throughput implications of IEEE 802.11 MAC, Proceedings IEEE International Performance, Computing, and Communications Conference, 2002, pp. 281–290.

O. S. Jarquin, Multi-hop radio networks in rough terrain: Some traffic-sensitive MAC algorithms, in Keshavarzian et al., Energy-efficient link assessment in wireless sensor networks, Proceedings InfoCom, 2004.

J. Li et al., Performance evaluation of modified IEEE 802.11 MAC for multi-channel multi-hop ad hoc networks, *J. Interconnection Networks*, 4(30):345–359, 2003.

Nilsson, Performance analysis of traffic load and node density in ad hoc networks, Proceedings European Wireless, 2004.

W. T. Raisinghani et al., Improving TCP performance over mobile wireless environments using cross-layer feedback, Proceedings IEEE International Conference on Personal Wireless Communications, 2002, pp. 81–85.

B. Raman et al., Arguments for cross-layer optimizations in Bluetooth scatternets, Proceedings Symposium on Applications and the Internet, 2001, pp. 176–184.

S. Shakkottai et al., Cross-layer design for wireless networks, *IEEE Communications*, Oct. 2003, pp. 74–80.

S. Toumpis and A. J. Goldsmith, Performance, optimization, and cross-layer design of media access protocols for wireless ad hoc networks, Proceedings IEEE ICC, 2003.

X. Yu, Improving TCP performance over mobile ad hoc networks by exploiting cross-layer information awareness, Proceedings MobiCom, 2004.

W. Yuen et al., A simple but effective cross-layer networking system for mobile ad hoc networks, Proceedings IEEE PIMRC, 2002.

M. Zuniga and B. Krishnamachari, Analyzing the transitional region in low power wireless links, Proceedings SECON, 2004.

Chapter 10

Applications and Recent Developments in Ad Hoc Networks

10.1 Introduction

Numerous factors associated with technology, business, regulation, and social behavior naturally and logically speak in favor of wireless ad hoc networking. Mobile wireless data communication, which is advancing in terms of both technology and usage penetration, is a driving force, thanks to the Internet and the success of second-generation cellular systems. As we look to the horizon, we can finally glimpse a view of truly ubiquitous computing and communication. In the near future, the role and capabilities of short-range data transaction are expected to grow, serving as a complement to traditional large-scale communication: most machine communication as well as oral communication between human beings occur at distances of less than ten meters; also, as a result of this communication, the two communicating parties often have a need to exchange data. As an enabling factor, license-exempted frequency bands invite the use of developing radio technologies (such as Bluetooth) that admit effortless and inexpensive deployment of wireless communication. In terms of price, portability, and usability, and in the context of an ad hoc network, many computing and communication devices, such as Personal Digital Assistants (PDAs) and mobile phones, already possess the attributes that are desirable. As advances in technology continue, these attributes will be enhanced even further. Finally, we note that many mobile phones and other electronic devices already are or will soon be Bluetooth enabled. Consequently, the ground for building more complex ad hoc networks is being laid. In terms of market acceptance, the realization of a critical mass is certainly positive. But perhaps even more positive—as relates to the end user—is that consumers of Bluetooth-enabled devices obtain a lot of as-yet unraveled ad hoc functionality at virtually no cost. Ad hoc wireless networks are communication networks in which communication devices are mobile and are not tied in any fixed topological infrastructure. As the devices move, the network topology changes with the movement. The mobile devices in such networks not only are the source and destination of information being exchanged, but also act as intermediate devices to relay information from one device to another, for those who are not within the communication range of each other. These networks have dynamic topology, bandwidth-constrained variable-capacity wireless links, energy-constrained operation, and limited physical security. These networks have tremendous potential for commercial and military applications, and are particularly useful for providing communication support where no communication infrastructure exists or where the deployment of a fixed infrastructure is economically not feasible. Their potential applications include military operations, emergency situations, health care, home networking, academic settings, and disaster recovery operations. For meeting the need for fast and reliable information exchange, communications networks have become an integral part of our society. Recent technological advancements in information technology and continued miniaturization of mobile communication devices have increased the use of wireless communication environments manifold. As a result of this tremendous growth, wireless networks have witnessed rapid changes and the development of new applications. One such change is the development of mobile ad hoc wireless networks. Wireless ad hoc networks are essentially communication networks in which communication devices are mobile and are not tied in any fixed topological infrastructure. They are self-organizing and adaptive. As the devices move, the network topology changes with the movement. The devices in such networks not only are the source and destination of information being exchanged, but also act as intermediate devices to relay information from one device to another, for those who are not within the communication range of each other. These networks are characterized by dynamic topologies, bandwidth-constrained variable-capacity wireless links, energy-constrained operation, and limited physical security. Ad hoc networks have tremendous potential in commercial and military applications. These networks are particularly useful for providing communication support where no fixed infrastructure exists or where the deployment of a fixed infrastructure is economically not feasible. Examples of disaster situations include earthquake and flooding where the rescue teams need to coordinate themselves without the availability of fixed networks, military operations where communication is in a hostile environment, businesses where employees share information in a conference, educational settings where students using laptop computers participate in an interactive lecture, and many other similar situations. These networks are established to meet a temporary networking need for a specific duration of time, and when the need disappears so do these networks. Ensuring effective communication among the nodes is one of the major challenges in ad hoc networks. Because of the limited range of each host's wireless transmission, to communicate with hosts outside its transmission range, a host needs to act as a relay station in forwarding packets to the destination. Therefore, some form of routing protocol, which can address a diverse range of issues such as low bandwidth, mobility, and low power consumption, is necessary in ad hoc networks. The mobile devices work together in a dynamic but cooperative environment to maintain the communication channels. It is possible for the mobile devices in an ad hoc wireless network to drift away from each other and not be able to maintain communication with others. In such a situation, an ad hoc network may be divided into two or more independent ad hoc networks. On the other hand, it is also possible that mobile devices in two or more ad hoc networks come in close proximity of each other and fuse into one larger ad hoc network. One can imagine the challenges of managing such a dynamic communication environment. Another form of ad hoc wireless networks is sensor networks. In sensor networks, the mobile devices are very small in size. These devices could be as small as a grain of rice and are self-sufficient in all respects—transmitting, receiving, processing, and power. These sensors can be programmed to suit any given application.

10.2 Typical Applications

Mobile ad hoc networks have been the focus of many recent research and development efforts. So far, ad hoc packet-radio networks have mainly been considered for military applications, where a decentralized network configuration is an operative advantage or even a necessity. In the commercial sector, equipment for wireless, mobile computing has not been available at a price attractive to large markets. However, as the capacity of mobile computers increases steadily, the need for unlimited networking is also expected to rise. Commercial ad hoc networks could be used in situations where no infrastructure (fixed or cellular) is available. Examples include rescue operations in remote areas, or when local coverage must be deployed quickly at a remote construction site. Ad hoc networking could also serve as wireless public access in urban areas, providing quick deployment and extended coverage. The access points in networks of this kind could serve as stationary radio relay stations that perform ad hoc routing among themselves and between user nodes. Some of the access points would also provide gateways via which users might connect to a fixed backbone network. At the local level, ad hoc networks that link notebook or palmtop computers could be used to spread and share information among participants at a conference. They might also be appropriate for application in home networks where devices can communicate directly to exchange information, such as audio-video devices, alarms, and configuration updates. Perhaps the most far-reaching applications in this context are more or less autonomous networks of interconnected home robots that clean, do dishes, mow the lawn, perform security surveillance, and so on. Some people have even proposed ad hoc multihop networks (denoted sensor networks)–for example, for environmental monitoring, where the networks could be used to forecast water pollution or to provide early warning of an approaching tsunami. Short-range ad hoc networks can simplify intercommunication between various mobile devices (such as a cellular phone and a PDA) by forming a personal area network (PAN), and thereby eliminate the tedious need for cables. This could also extend the mobility provided by the fixed network (that is, mobile Internet Protocol, or IP) to nodes further out in an ad hoc network domain. The Bluetooth system is perhaps the most promising technology in the context of personal area networking.

10.2.1 Personal Area Network (PAN)

A network extension seen from the viewpoint of the traditional mobile network, a Bluetooth-based PAN opens up a new way of extending mobile networks into the user domain. Someone on a trip who has access to a Bluetooth PAN could use the General Packet Radio Service/Universal Mobile Telecommunication System (GPRS/UMTS) mobile phone as a gateway to the Internet or to a corporate IP network. In terms of traffic load in the network, the aggregate traffic of the PAN would typically exceed that of the mobile phone. In addition, if Bluetooth PANs could be interconnected with scatternets, this capacity would be increased. Figure 10.1 shows a scenario in which four Bluetooth PANs are used.

The PANs are interconnected via laptop computers with Bluetooth links. In addition, two of the PANs are connected to an IP backbone network, one via a local area network (LAN) access point and the other via a single GPRS/UMTS phone. A PAN can also encompass several different access technologies distributed among its member devices, which exploit the ad hoc functionality in the PAN. For instance, a notebook computer could have a wireless LAN (WLAN) interface (such as Institute of Electrical and Electronic Engineers [IEEE] 802.11 standards or HiperLAN/2) that provides network access when the computer is used indoors. Thus, the PAN would benefit from the total aggregate of all access technologies residing in the PAN devices. As the PAN concept matures, it will allow new devices and new access technologies to be incorporated into the PAN framework. It should also eliminate the need to create hybrid devices, such as a PDA–mobile phone combination, because the PAN network will instead allow for wireless integration. In other words, it will not be necessary to trade off form for function. In all the scenarios discussed above, it should be emphasized that close-range radio technology, such as Bluetooth, is a key enabler for introducing the flexibility represented by the PAN concept.

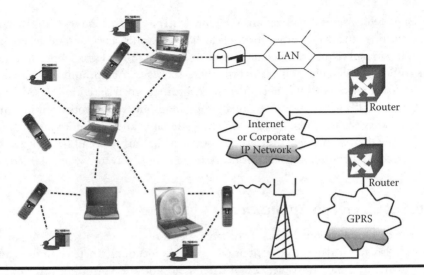

Figure 10.1 Personal area network (PAN) scenario with four interconnected PANs, two of which have an Internet connection via a Bluetooth local area network (LAN) access point and a GPRS/UMTS phone.

10.3 Applications and Opportunities

Ad hoc wireless networks can be used wherever there is a need for establishing a networking environment for a limited duration of time. These networks provide tremendous opportunities and can be used in numerous situations, particularly where a communication infrastructure is nonexistent or difficult to establish within timing constraints. Typically, such applications include the following:

- Search-and-rescue applications in disaster situations.
- Defense (army, navy, and air force) applications.
- Health care applications.
- Academic environment applications.
- Industrial or corporate environment applications.

There are many other applications that can utilize ad hoc wireless networks.

10.3.1 Search-and-Rescue Applications

When we face an unfortunate situation such as an earthquake, hurricane, or similar disaster, ad hoc wireless networks can prove to be very useful in search-and-rescue operations. In general, disasters leave a large population without power and communication capabilities for they destroy the infrastructures. Ad hoc wireless networks can be established without such infrastructures and can provide communications among various relief organizations for coordinating their rescue operations. Wireless sensor networks (another form of ad hoc networks) can be used to conduct searches for survivors and provide care in a timely manner. Rescue operations also use robots in searching for survivors. These robots can communicate with each other using wireless ad hoc networks and coordinate their activities. Based on the size of area affected by a disaster, an appropriate number of robots (forming an ad hoc network) can be deployed for searching the area and for information gathering in the shortest possible time. The information so gathered can be analyzed and processed, and appropriate relief or help can be readily directed where needed.

10.3.2 Defense Applications

Secure communications is one of the key aspects of any successful defense operations. Also, many defense operations take place in locations where communication infrastructure is not available. Use of wireless ad hoc and sensor networks in such situations becomes very useful and expedient. Different units (army, navy, and air force) involved in defense operations also need to maintain communication with each other. Air force planes flying in a formation may establish an ad hoc wireless network for communicating with each other and for sharing images and data among themselves. Army groups on the move can also use ad hoc wireless networks for communicating among themselves. The same applies to navy personnel. A nice feature of such a communication environment is that the ad hoc network moves with you as individuals move or planes fly. One of the many applications of ad hoc wireless (particularly sensor) networks is information gathering. Intelligence gathering for defense purposes can make use of sensor networks very effectively. The sensors used for such applications are essentially disposable and are used for an application once. The sensors can be deployed in large quantities over a selected area chosen for intelligence gathering. The sensors can be deployed by air or by other appropriate means. Because of their tiny size, these sensors will remain suspended in the air for some time. During that time, they can collect information that they have been programmed for, process the information, share among other nearby sensors, reach a consensus, and transmit information to a central location. The information can then be analyzed at the central processing facility, and a decision about the next step can be made. Sensor networks can also be used for tracking objects or targets, which is one of the critical applications in defense settings. With rapid advancements in semiconductor technologies, size of the electronic devices is becoming smaller and smaller. At the same time these devices are able to muster higher and higher processing power on tiny chips. These advancements have led to the development of wearable computers. The idea of wearable computers is not that new, but the idea of a smart dress (that consists of many tiny computers, or sensors) is relatively recent. In a smart dress, tiny computers are connected by tiny wires or by wireless means, which can exchange information with each other, process information, and take an action that they are programmed to do. A smart dress may be programmed to monitor certain conditions and vital signs of an individual on a regular basis. This could become very useful for defense personnel in combat situations. The monitored information can be processed, and appropriate action can be taken by the dress, if needed. A smart dress may even be able to indicate the exact location of the problem. It may also be able to call for help if seriousness of the situation warrants that.

10.3.3 *Health Care Applications*

Exchanging multimedia (audio, video, and data) information between a patient and health care facilities is very helpful in critical and emergency situations. An individual who is being transported to a hospital by an ambulance may exchange information using ad hoc communication networks. A health care professional, in many situations, is in a much better position to diagnose and prepare a treatment plan for an individual if he or she has video information rather than just audio or data information. For instance, video information may be helpful in assessing the reflexes and viewing the coordination capability of a patient. Similarly, the level of injuries of a patient can be established better with visual information than with just audio or other descriptive information. Real-time ultrasound scans of a patient's kidneys, heart, or other organs may be very helpful in preparing a treatment plan for a patient who is being transported to a hospital, prior to his or her arrival in the hospital. Such information can be transmitted through wireless communication networks, from an ambulance to a hospital or to other health care professionals who are currently scattered at different places but are converging toward the hospital for treating the patient being transported. Ad hoc wireless networks established within a (smart) home can also be very useful for monitoring homebound patients. Such homes may be able to make some basic decisions (based on information exchanged between various sensors participating in an ad hoc network) that are beneficial to the elderly population. Some of the actions that smart homes can take include monitoring the movement patterns inside a home, recognizing a fall of a human being, recognizing an unusual situation, and informing a relevant agency so that appropriate help can be provided, if needed. The concept of a smart dress, discussed in the subsection on defense applications (above), can also be used to monitor health conditions of patients. Such dresses may become very useful for providing health care for our elderly population.

10.3.4 *Academic Environment Applications*

Most of the academic institutions either already have wireless communication networks or are in the process of establishing such facilities. Such an environment provides students and faculty a convenient environment to interact and accomplish their mission. Ad hoc wireless networks can enhance such an environment and add many attractive features. For instance, an ad hoc wireless communication network can be established among the instructor and the students enrolled in his/her class. Such a setting can provide an easy and convenient mechanism for instructor to distribute handouts to all the students in the class and also for students to submit their assignments. Sharing information among the class participants can be as easy as click of a key on the keyboard. Due to the aura of mobility attached with the ad hoc wireless networks, such networks can also be established while on a field trip and industrial visits. Staying in touch cannot be any easier than this.

10.3.5 Industrial Environment Applications

Most industrial or corporate sites have wireless communication networks in place, particularly in manufacturing environments. Manufacturing facilities, in general, have numerous electronic devices that are interconnected. Having wired connectivity leads to cluttering and crowding of space, which not only pose safety hazards but also adversely affect reliability. Use of wireless communication networks eliminates many of these concerns. If the connectivity is in the form of ad hoc wireless communication networks, that adds many attractive aspects, including mobility. The devices can be easily relocated, and the networks reconfigured based on the requirements as they arise. At the same time, communication among various communicating entities can be maintained, and corporate meetings can take place without employees gathering in the same room.

10.4 Challenges

Although ad hoc wireless communication networks represent technological marvels, there are many challenges that need to be addressed for fully harvesting their benefits. As with all mobile communication environments, ad hoc wireless communications operate with the following constraints:

- Limited communication bandwidth and capacity
- Limited battery power and life
- Size of the mobile devices
- Information security
- Communication overhead

Wireless communication operates with limited bandwidth, which implies that only a limited amount of information can be transmitted over a period of time. Efficient transmission techniques will pave the way for increased capacity. However, that is not enough, and innovative approaches for optimal use of available bandwidth and capacity are needed. The concept of cellular communication structures and the use of transmission techniques such as code division multiple access (CDMA) are very helpful. Additional research is still needed to provide more efficient mechanisms for using the available communication bandwidth in a wireless communication environment. Mobile communication devices do not have access to unlimited power. Mobile devices use batteries, and batteries have a limited supply of power. The higher the power usage, the shorter the battery life will be. Efforts are being made to design devices that consume less power and adjust the strength of communication signals based on the distance between communicating points. In addition, efficient signal-processing techniques and algorithms are being developed that will require less power usage. With the advancements in semiconductor technologies, more and more electronic components can be placed on smaller chips. That has led to the development of mobile devices that are more powerful and less power hungry. As the size of these mobile devices continues to shrink, more and more features and functionalities are being added to these devices without much demand for power. The challenge is to maintain that trend. Wireless communication environments are more prone to security risks than others, and ad hoc wireless networks are no exception. Any level of desired information security can be achieved, but this adds additional (processing) overhead and requires additional bandwidth for transmission. Researchers are working on discovering mechanisms that will provide secure information transfer and at the same time will not add prohibitive overhead. Reducing the communication overhead for transferring information in ad hoc wireless communication networks is one of the biggest and most formidable challenges. When information needs to be transmitted from one device to another, a route or path needs to be established for exchanging information. In addition, some procedure for sharing the common pool of resources such as bandwidth has to be established. Bluetooth technology and IEEE 802.11 protocols provide mechanisms for sharing the resources. Also, there are several routing mechanisms that have been proposed for establishing a route between two communicating entities. The challenge that ad hoc wireless networks pose is that they have dynamic topology. To establish a route between two communicating devices, the network components need to be aware of the location of other devices. To make things complicated, devices are mobile and keep changing their locations. The procedures for establishing routes have to be dynamic and adaptive. The route that is established at the start of information transfer between two devices may not be the same when the information reaches its destination. Therefore, routing information needs to be as current as possible all the time. There are several possibilities about establishing and maintaining routes. Routes can be established on a proactive basis or an on-demand basis. The procedures that establish routes on a proactive basis incur more overhead because establishment of all routes may not be necessary. If routes are established on a proactive basis and frequently enough, they will be not only current but also immediately available to any communication device that needs to send information to some other device. On the other hand, if routes are established on an on-demand basis, the overhead incurred will be less because only those routes will be established that are needed. However, an on-demand routing mechanism will introduce more waiting for devices because they will need to wait for the route to be established before communication of information can begin. There are many hybrid routing mechanisms that have been proposed. However, the challenge for establishing the best possible routes with the least possible overhead remains open.

10.4.1 Security

The research on mobile ad hoc network (MANET) security is still in its early stage. The existing proposals are typically attack oriented in that they first identify several security threats and then enhance the existing protocol or propose a new protocol to thwart such threats. Because the solutions are designed explicitly with certain attack models in mind, they work well in the presence of designated attacks but may collapse under unanticipated attacks. Therefore, a more ambitious goal for ad hoc network security is to develop a multifence security solution that is embedded into possibly every component in the network, resulting in depth protection that offers multiple lines of defense against many both known and unknown security threats. This new design perspective is called "resiliency-oriented security design." The resiliency-oriented security solution possesses several features. First, the solution seeks to attack a bigger problem space. It attempts not only to thwart malicious attacks but also to cope with other network faults due to node misconfiguration, extreme network overload, or operational failures. In some sense, all such faults, whether incurred by attacks or misconfigurations, share some common symptoms from both the network and end-user perspectives and should be handled by the system. Second, resiliency-oriented design takes a paradigm shift from conventional intrusion prevention to intrusion tolerance. In a sense, certain degrees of intrusions or compromised or captured nodes are the reality to face, not the problem to get rid of, in MANET security. The overall system has to be robust against the breakdown of any individual fence, but its performance does not critically depend on a single fence. Even though attackers intrude through an individual fence, the system still functions, but possibly with graceful performance degradation. Third, as far as the solution space is concerned, cryptography-based techniques just offer a subset of toolkits in a resiliency-oriented design. The solution also uses other noncryptography-based schemes to ensure resiliency. For example, it may piggyback more "protocol-invariant" information in the protocol messages, so that all nodes participating in the message exchanges can verify such information. The system may also exploit the rich connectivity of the network topology to detect inconsistency of the protocol operations. In many cases, routing messages are typically propagated through multiple paths, and redundant copies of such messages can be used by downstream nodes. Fourth, the solution should be able to handle unexpected faults to some extent. One possible approach worth exploring is to strengthen the correct operation mode of the network by enhancing more redundancy at the protocol and system levels. At each step of the protocol operation, the design makes sure what it has done is completely along the right track. Anything deviating from valid operations is treated with caution. Whenever an inconsistent operation is detected, the system can raise a suspicion flag and query the identified source for further verification. This way, the protocol tells right from wrong because it knows right with higher confidence, not necessarily knowing what is exactly wrong. The design strengthens the correct operations and may handle even unanticipated threats in runtime operations. Next, the solution may also take a collaborative security approach, which relies on multiple nodes in a MANET to provide any security primitives. Therefore, no single node is fully trusted. Instead, only a group of nodes will be trusted collectively. The group of nodes can be nodes in a local network neighborhood or all nodes along the forwarding path. Fifth, the solution relies on multiple fences, spanning different devices, different layers in the protocol stack, and different solution techniques, to guard the entire system. Each fence has all the functional elements of prevention, detection and verification, and reaction. The above mentioned resiliency-oriented MANET security solution poses grand yet exciting research challenges. How to build an efficient fence that accommodates each device's resource constraint poses an interesting challenge. Device heterogeneity is one important concern that has been largely neglected in the current security design process. However, multifence security protection is deployed throughout the network, and each individual fence adopted by a single node may have different security strength due to its resource constraints. A node has to properly select security mechanisms that fit well into its own available resources, deployment cost, and other complexity concerns. The security solution should not stipulate the minimum requirement a component must have. Instead, it expects the best effort from each component. The more powerful a component is, the higher degree of security or resiliency it has. Next, identifying the system principles of how to build such a new generation of network protocols remains unexplored. The state-of-the-art network protocols are all designed for functionality only. The protocol specification fundamentally assumes a fully trusted and well-behaved network setting for all message exchanges and protocol operations. It does not anticipate any faulty signals or ill-behaved nodes. We need to identify new principles to build the next-generation network protocols that are resilient to faults. There exist only a few piecemeal individual efforts. Finally, evaluating the multifence security design also offers new research opportunities. The effectiveness of each fence and the minimal number of fences the system has to possess to ensure some degree of security assurances should be evaluated through a combination of analysis, simulations, and measurements in principle. However, it is recognized that the current evaluation for state-of-the-art wireless security solutions is quite ad hoc. The community still lacks effective analytical tools, particularly in a large-scale wireless network setting. The multidimensional trade-offs among security strength, communication overhead, computation complexity, energy consumption, and scalability still remain largely unexplored. Developing effective evaluation methodology and toolkits will probably need interdisciplinary efforts from research communities working in wireless networking, mobile systems, and cryptography. Evaluating ideas and algorithms in real-world mobile ad hoc networks is difficult because it is hard to achieve repeatable and reliable results. The most important problem with these evaluations is the wireless channel due to interference, diffraction, and changing atmospheric conditions. Another problem is the total number of nodes involved in creating larger ad hoc networks. Hence, simulators like NS-2, OPNET, or others are used to analyze ad hoc networks in a sealed environment. But even here, problems with realistic radio propagation and user mobility models remain.

10.5 Highlights of the Most Recent Developments in the Field

The availability of cheaper, faster, and more reliable electronic components has stimulated important advances in computing and communication technologies. Theoretical and algorithmic approaches that address key issues in sensor networks, ad hoc wireless networks, and peer-to-peer networks—simply called SAP networks—play a central role in the development of emerging network paradigms. Filling the need for a comprehensive reference on recent developments, what are the central technical issues in these SAP networks? What are the possible solutions or tools available to address these issues?

Both coverage and wireless sensor networks are intrinsically multidisciplinary research topics. Therefore, a wide body of scientific and technological work is related to research presented in this chapter. In this section, we briefly cover only the most directly related areas: sensors, wireless ad hoc sensor networks, the coverage problem, and related sensor network problems such as location discovery and deployment.

10.5.1 Sensors

A sensor is a device that produces a measurable response to a change in a physical condition, such as temperature or magnetic field. Although sensors have been around for a long time, two recent technological revolutions have greatly enhanced their importance and their range of application. The first was the connection of sensors to computer systems, and the second was the emergence of microelectromechanical system (MEMS) sensors with their small size, small cost, and high reliability. There are a number of comprehensive surveys for a variety of sensor systems.

10.5.2 Wireless Ad Hoc Sensor Networks

Recently, wireless sensor networks have been attracting a great deal of attention commercially and in research. In particular, the practical emergence of wireless ad hoc networks is widely considered revolutionary in terms of being both a paradigm shift as well as an enabler of new applications. In ad hoc networks there is no fixed network infrastructure (such as in cellular phone networks), and therefore they can be deployed and adapted much more rapidly. Furthermore, the integration of inexpensive, power-efficient, and reliable sensors in nodes of wireless ad hoc networks, with significant computational and communication resources, opens new research and engineering vistas. Applications range from connecting the Internet to the physical world to creating new proactive environments. At the same time, wireless sensor networks pose a number of demanding new technical problems, including the need for new Digital Signal Processing (DSP) algorithms, operating systems, low power designs, and integration with biological systems.

10.6 Summary

The development of ad hoc wireless networks and sensor networks provides tremendous opportunities in many areas including disaster recovery, defense, health care, academic, and industrial environments. However, there are many challenges that need to be addressed as well. The challenges include developing the following: mechanisms for efficient use of limited bandwidth and communication capacity, mechanisms for reducing power consumption and hence extending the battery life, smaller but more powerful mobile devices, algorithms for enhancing information security, and efficient routing procedures. These are major challenges to overcome, but steady progress is being made to address these.

10.7 Problems

10.1 List the important applications of wireless networks.
10.2 Describe PAN with a suitable example.
10.3 Explain the typical applications of ad hoc networks.
10.4 Discuss the challenges of ad hoc networks.
10.5 Highlight the opportunities in ad hoc networks.
10.6 Describe the security issues and challenges in mobile ad hoc networks.
10.7 Discuss the recent developments in ad hoc networks.

Index